Robert M Bancroft, Francis J Bancroft

Tall Chimney Construction

A practical treatise on the construction of tall chimney shafts

Robert M Bancroft, Francis J Bancroft

Tall Chimney Construction
A practical treatise on the construction of tall chimney shafts

ISBN/EAN: 9783337105792

Printed in Europe, USA, Canada, Australia, Japan

Cover: Foto ©berggeist007 / pixelio.de

More available books at **www.hansebooks.com**

TALL CHIMNEY CONSTRUCTION.

A

𝔓ractical 𝔗reatise on the 𝔠onstruction

OF

TALL CHIMNEY SHAFTS,

CONTAINING

DETAILS OF UPWARDS OF EIGHTY EXISTING MILL, ENGINE-
HOUSE, BRICK WORKS, CEMENT WORKS, AND OTHER CHIMNEYS
IN ENGLAND, AMERICA AND THE CONTINENT,

CONSTRUCTED IN

BRICK, STONE, IRON AND CONCRETE.

BY

ROBERT M. BANCROFT

(Past President Civil and Mechanical Engineers' Society, London, England),

AND

FRANCIS J. BANCROFT

(Assistant Municipal Surveyor).

───────◆───────

MANCHESTER:
JOHN CALVERT, 99, GREAT JACKSON STREET, HULME, AND 100, KING STREET.

LEWES:
FARNCOMBE AND CO., "EAST SUSSEX NEWS" OFFICES.

1885.

JOHN S. PRELL
Civil & Mechanical Engineer.
SAN FRANCISCO, CAL.

LEWES
FARNCOMBE AND CO.,
PRINTERS.

THE Authors wish to state that it has been their aim throughout to place before the reader a practical treatise, and for that purpose have collected information respecting upwards of eighty chimney shafts—brick, stone, iron and concrete—from various parts of Great Britain, the Continent and America, and they take this opportunity of thanking their numerous friends for the uniform courtesy with which their enquiries have been answered.

In January, 1878, a paper on Chimney Construction was read by Mr. R. M. BANCROFT, Past President, before the Civil and Mechanical Engineers' Society, and the interest with which it was received led him to further his investigations. In December, 1883, a paper on the same subject was again read before the Civil and Mechanical Engineers' Society, by the Authors, the information and examples having been greatly increased. Since that date the work has been added to, revised and further illustrated.

ILLUSTRATIONS.

INDEX.

TALL CHIMNEY CONSTRUCTION.

References : W. J. M. RANKINE, R. WILSON, THOMAS BOX, R. ARMSTRONG, HOLYROYD SMITH, P. CARMICHAEL, R. ANDERSON, GRAHAM SMITH, D. KIRKALDY, H. FAIJA, J. WAUGH.

CHIMNEYS are constructed principally for two purposes. Firstly, to create the necessary draught for the combustion of fuel ; Secondly, to convey the noxious gases to such a height that they shall be so intermingled with the atmosphere as not to be injurious to health.

A chimney shaft, when in work, contains a tall column of heated air, which, being lighter than the outside atmosphere, is forced upward by a corresponding column of atmospheric air pressing into the entrance of the furnace ; thus a displacement of hot air is constantly being effected, and its place filled by normal air forcing itself through the furnace of the boiler, which air is in its turn heated and displaced. The column of atmospheric air and the column of rarefied air in the chimney are somewhat like a pair of scales, or the two ends of a lever of which the boiler is the fulcrum.

Foundations.—In building large chimneys one of the most important points is the construction of the foundation. Very much will depend, of course, upon the nature of the ground. When on solid rock, it is only necessary to excavate to such a depth that the heat of the gases will not materially affect the natural stone, and to a depth sufficient to allow the necessary spreading of the base. In many instances, however, chimney stacks have to be built near rivers and on sites where the upper strata are of alluvial clay or made ground, and it is necessary to carry the foundation deep down until a stiff clay, hard sand, or rock bottom is reached. This frequently entails excavation 25′ or 30′ deep or even more, and it is not only requisite that the foundation should be large enough to carry the superincumbent weight, but also that it should be of such an area

B

that it will not allow the base to be forced into the yielding-
ground. These deep foundations are usually constructed or
concrete.   In some cases piles are driven in to form the foundation, as, among others mentioned further on, in a brick chimney
erected at Boston, England, and in an iron chimney constructed
at Ohio, U.S.A.   This piling is a measure on which the engineer
must decide upon the advisability of using, so as to economise
material without risking unequal subsidence, which cannot be
too carefully guarded against; and; in fact, it is the practice
in the erection of tall stacks to construct the foundation and
pedestal, if any, and allow them to stand some considerable
time before proceeding with the shaft proper, in order that the
work may set, and any slight settling take place, before a great
weight is built upon it.   As a remarkable instance of the general
settlement of the foundation of a shaft, we may mention a
chimney which was built by Mr. Clegg, at Fulham, over a
quicksand, in which an iron rod sank to a depth of 15' with
little more than its own weight as pressure.   During the erection
the concrete foundation sank bodily 1' 4½" without cracking
the shaft or causing it to deviate from the perpendicular.   From
this it will naturally follow that in all cases the ground at
the foundation should be equally resistant, or unequal settling
will take place, as in the disastrous case of the Newland's
Mill chimney, Bradford, as hereinafter detailed.   Some of the
pressures exerted upon the foundations are given under the
respective descriptions of the chimneys, most of the other
pressures can be worked out from the data supplied.

Weather.—Shafts should be erected in the summer months;
on no account should the work be proceeded with in frosty
weather.

Progress.—Shafts should be constructed at the rate of from
2' to 2½' in height per day, but of course the progress is
largely dependent upon the size of the shaft being erected; the
taller the shaft the more care should be exercised in allowing
the mortar to set, and the foundations to gradually take and
settle down to the weight of the superincumbent mass.   In
Lancashire large shafts are built about half their ultimate height,
and then left six months to consolidate before completion.

Brickwork, &c.—The bricks used should be picked stocks,
hard and sound, with square sharp edges, thoroughly burnt and

of uniform thickness. They should be well wetted with water before being laid.

The joints of the brickwork should be well flushed up with mortar every course; this is much to be preferred to grouting every 2nd or 3rd course. The grouting being fluid mortar becomes porous and possesses little adhesive power, as the water evaporates. Grouting should, therefore, be discountenanced.

The brickwork should be laid in mortar for the most part because cement is destroyed by a strong heat. Any $4\frac{1}{2}''$ work at the top should, however, be constructed in good cement. With so thin a wall the heat is rapidly carried off by the external air, and in such a case the cement will be uninjured.

Ordinary stock bricks will withstand a heat of 600°; where a heat is anticipated greater than this, fire-bricks should be used.

Bond.—In large factory chimney shafts the longitudinal tenacity which resists any force tending to split the chimney is of more importance than the transverse tenacity; therefore, in these structures, it is advisable to have, say, three or four courses of stretchers to one course of headers.

In some circular stacks an uniform header bond is adopted for the outside courses of the brickwork. This is a practice condemned by some authorities, but is almost unavoidable in the construction of small circular shafts, unless purposely made bricks are used, owing to the sharp curvature of the work; in shafts having large diameters it should never be employed.

The longitudinal strength of a shaft is much increased by building in hoop-iron every few courses, and is a practice often adopted and to be commended. Care must, however, be taken to fix a good lightning conductor to a shaft thus constructed, as the stack would form a great attraction to the electric fluid.

Expansion.—Chimney shafts should not be tied to any other work or buildings, and should have no woodwork or anything fixed to them, on account of the settlement that takes place after the shaft has been erected, the expansion caused by the heated gases, and the oscillations caused by wind.

Supervision.—The shafts should be plumbed and levelled every 3', or oftener, so as to obtain a regular batter, and keep the stack erect; and care should be taken that in angular shafts the quoins are built without twist.

Caps, Copings and Cornices.—The stone coping or cornice of a chimney will seldom require more to hold it together than two good cramps across each joint; they should be of copper, or double dove-tailed slate dowels. On no account should iron cramps be used, as they will oxidise and burst the stone. Heavy and large caps are often the source of great danger, inconvenience and expense (see Brooks & Son's Chimney, page 48), as the cap at top in a gale of wind acts upon the shaft as a weight at the end of a long lever. The cap, when finished, should be a complete whole, or so bound together that the joints cannot open, and be so proportioned that the centres of gravity, of its respective component parts, all fall within the outer circle of the shaft on which they rest, and the cap should be designed so that the wind striking against it is deflected upwards.

Minimum Height.—The minimum height of chimney shafts allowed by many Town Improvement Acts, as in Manchester, Bradford, Leeds and other towns, is 90'.

FURNACE CHIMNEY SHAFTS.

Every chimney shaft, for the furnace of a steam boiler, brewery, distillery, or manufactory, shall be carried up throughout in brickwork and mortar, or cement, of the best quality.

Every furnace chimney shall be built upon a bed of concrete, to the satisfaction of the District Surveyor.

The base of the shaft shall be solid to the top of the footings, and the footings shall spread equally all round the base by regular offsets to a projection on both sides equal to the thickness of the wall at the base.

The width, measured externally, of a furnace chimney shaft at the base, or at that portion immediately above the footings, shall be as follows :—

If *square* on plan, at least one-tenth of the total height of the shaft.

If *octagonal* on plan, at least one-eleventh of the total height of the shaft.

If *circular* on plan, at least one-twelfth of the total height of the shaft.

Every furnace chimney shaft shall have a batter of $2\frac{1}{2}''$ at least in every 10' of height, or 1 in 48.

The brickwork shall be at least $8\frac{1}{2}''$ in thickness at the top of the shaft and for not exceeding 20' below, and shall be increased at least $4\frac{1}{2}''$ in thickness for every 20' of additional height measured downwards.

No portion of the walls of a furnace chimney shaft shall be constructed of fire-brick, and any fire-brick lining to be used must be in addition to the thickness of, and independent of, the brickwork.

Every cap, cornice, pedestal, string-course, or other variation from plain brickwork, shall be in addition to the thickness of brickwork prescribed by the foregoing rules, and no cornice shall project more than the thickness of the brickwork at the top of the shaft.

TESTS AND STRENGTHS OF MATERIALS.

(Except where otherwise stated, made by D. Kirkaldy, London.)

———◆———

BRICKS.

———

Mr. J. C. Edwards' Brick Works, Ruabon.

———

The following are results of experiments to ascertain the resistance, to a gradually increased thrusting stress, of twelve bricks manufactured at these works :—

Description.	Dimensions, inches.	Base area, square inches	Stress in pounds when		Crushed, steel-yard dropped.
			Cracked slightly.	Cracked generally.	
Red Brick (no recess)	3·10 . 8·75 × 4·28	37·45	336,050	446,700	461,500
,, ,, ,,	3·13 . 8·75 × 4·26	37·27	225,300	391,500	455,100
,, ,, ,,	3·10 . 8·80 × 4·30	37·84	358,500	408,100	442,950
,, ,, ,,	3·19 . 8·78 × 4·26	37·40	215,100	236,250	376,900
,, ,, ,,	3·10 . 8·75 × 4·28	37·45	232,500	286,100	358,600
,, ,, ,,	3·15 . 8·73 × 4·29	37·45	168,250	223,500	271,300
Mean		37·47	255,950	333,025	394,391
Lbs. per square inch . .			6,830	8,887	10,525
Tons per square foot . .			439·2	571·5	676·8
Blue Bricks (no recess)	3·02 . 8·99 × 4·37	39·28	307,100	385,100	388,050
,, ,, ,,	3·00 . 8·97 × 4·34	38·92	261,900	338,100	377,030
,, ,, ,,	3·03 . 9·00 × 4·37	39·33	148,300	246,250	344,100
,, ,, ,,	3·05 . 9·00 × 4·38	39·42	181,800	251,500	327,700
,, ,, ,,	3·04 . 8·97 × 4·33	38·84	153,000	286,100	310,100
,, ,, ,,	3·01 . 8·95 × 4·36	39·02	258,600	300,200	302,050
Mean		39.13	218,450	301,208	341,505
Lbs. per square inch . .			5,582	7,697	8,727
Tons per square foot . .			358·9	494·9	561·2

Bedded between pieces of pine ¾″ thick.

HIGH BROOMS BRICK COMPANY, TUNBRIDGE WELLS.

The following table gives the average pressure in pounds on a single brick, each result being the mean of six tests:—

Description.	Base area square inches	Slightly cracked. lbs.	Generally cracked. lbs.	Crush'd, steel yard dropped. lbs.	Tons per square foot.
Sewage Bricks (dark) . .	39·87	170,847	226,853	253,398	408·7
Do. do. (light) . .	39·99	136,933	209,905	232,863	374·5
Brindle do. (dark) . .	36·97	162,497	211,190	296,790	516·2
Do. do. (light) . .	36·52	94,922	141,688	235,585	414·8
Blue do.	40·64	188,125	250,950	332,043	525·4
Wire-cut do.	40·80	105,658	157,927	232,337	366·2
Common Building . . .	37·94	74,987	97,070	118,825	201·4

The samples were bedded between pieces of pine ⅜" thick.

STAFFORDSHIRE VITRIFIED BLUE BRICKS.

Size of bricks, 2·55" × 9·03" × 4·30".
Mean base area of six samples, 39·06 square inches.
All samples bedded between pieces of pine ¼" thick.

J. HAMBLETT, West Bromwich.	Stress in lbs. when		
	Cracked slightly.	Cracked generally.	Crushed, steel-yard dropped.
1st sample	187·5	341·6	449·68
2nd ,, ...	164·3	306·0	404·27
3rd ,, ...	153·0	291·0	396·16
4th ,, ...	148·2	285·4	387·82
5th ,, ...	124·6	244·1	382·75
6th ,, ...	115·8	216·2	366·34
Mean	148·9	280·7	397·83
Equal to lbs. per sq. inch .	3·812	7·186	10·185
,, Tons per sq. foot	245·1	462·1	654·9

The following table gives results of tests made during the erection of Blackfriars Bridge:—

BRICKS IN PIERS FOUR COURSES HIGH.

Description of bricks.	Size of pier in bricks.	Mortar.	Failing slightly. tons per foot super.	Entirely crushed tons per foot super.
Common Stock, recessed .	1½ X 1½	Lias Lime.	17	27
Do. do. . .	,, ,,	,,	21	30
Red bricks, machine made	,, ,,	,,	20	40
Do. hand made . .	,, ,,	,,	20	36
Gault	,, ,,	Roman Cement	24	59
Do.	1 X 1	,,	54	72
Clark's Sudbury Machine .	,, ,,	Portland.	49	76
Uxbridge red, hand made .	,, ,,	,,	44	53

D. W. BARKER'S BRICK WORKS, WORCESTER.

Samples of bricks used in the construction of chimney shaft 160' high at the above works:—

Description.	Dimensions in inches.	Base area in square inches	Compressive stress in pounds when		
			Cracked slightly.	Cracked generally.	Crush'd, steel yard dropped.
Pressed, recessed top and bottom	3·20 X 9·14 X 4·50	41·13	45,680	86,220	91,180
,, ,,	,, ,, ,,	,,	45,590	79,775	90,320
,, ,,	,, ,, ,,	,,	38,760	77,830	89,640
,, ,,	,, ,, ,,	,,	36,180	70,960	85,820
Mean			41,552	78,696	89,240
Lbs. per square inch			1,010	1,913	2,170
Tons ,, foot			65	123	139·5
Builders, recessed top and bottom	3·20 X 9.30 X 4.50	41·85	40,960	97,240	113,220
			39,280	95,270	106,530
			36,490	87,382	101,202
			33,540	81,180	95,840
Mean			36,490	87,382	101,202
Lbs. per square inch			872	2,038	2,418
Tons ,, foot			56·1	134·2	155·5

Bedded between pieces of pine ¼ thick and recesses filled with cement.

AMERICAN BRICKS.

Compression of Bricks, tested for new Pension Building, Washington, D.C., at Watertown Arsenal, Mass., December 16, 1882. Bricks tested between flat iron compression platforms. Compression faces of bricks ground flat.

Marks on bricks.	Sectional area.	Ultimate strength. Total lbs.	Lbs. per sq. in.	Bearings.	Remarks.
W. H. West & Bro., Red	4·00 × 8·50 = 34·00	324,500	9,540	Even	Cracking sounds heard at 183,000-lbs.
,, ,, ,, Arch	3·95 × 8·50 = 33·58	255,200	7,600	Uneven, required ·01-inch packing	Cracking sounds heard at 30,000-lbs. Fractures in sight at 80,000-lbs.
,, ,, ,, Press	4·20 × 8·50 = 35·70	231,000	6,470	Even	Cracking sounds heard at 125,000-lbs. Cracks in sight at 150,000-lbs.
Washington Brick Co., Red	4·10 × 8·47 = 34·73	296,200	8,530	,,	Cracks in sight at 32,000-lbs. at corner. At 130,000-lbs. specimen covered, no more cracks in sight.
,, ,, Arch	3·80 × 8·30 = 31·54	324,500	10,290	,,	Cracking sounds heard at 65,000-lbs. specimen covered. When load reached 150,000-lbs. no cracks in sight.
,, ,, Press	4·10 × 8·35 = 34·24	314,700	9,190	,,	Cracking sounds heard at 68,000-lbs. Began to crack along edge at 110,000-lbs.
Childs & Son, Red	4·15 × 8·40 = 34·86	211,000	6,050	,,	Began to flake along one corner at 78,000-lbs.
,, ,,	4·10 × 8·46 = 34·69	209,300	6,030	,,	,, ,, off at edges at 107,000-lbs.
,, ,,	4·10 × 8·45 = 34·65	232,000	6,700	·00½-inch packing under one corner	,, ,, off at corner at 140,000-lbs.
,, ,, Arch	3·70 × 8·10 = 29·97	203,700	6,800	Even	Cracks appeared generally at 85,000-lbs.
,, ,, Press	4·20 × 8·40 = 35·28	210,200	5,960	,,	Began to flake at one edge at 110,000-lbs.
Burns, Russell & Co., Press	4·30 × 8·58 = 36·89	249,000	6,750	,,	Cracking sounds at 128,000-lbs. Cracks in sight at 140,000-lbs.

(Signed)

JOHN G. BUTLER,

Captain of Ordnance, Commanding.

TESTS OF BRICK USED IN THE CONSTRUCTION OF THE SOUTH
GATE HOUSE NEW RESERVOIR, NEW YORK, U.S.A.

Tested in Hatfield's Hydraulic Press for testing building
materials, built by Messrs. Hoe & Co. The bricks were those
known as *hard bricks*, and made at the yard of Mr. William Call,
Haverstran, on the Hudson River:—

No.	Thick.	Wide.	Broad.	Square in. exposed to pressure.	How bedded.	Remarks.
1	2·30	3·52	4·40	15·488	Between two pieces of board ⅛-in. thick.	At 30,000-lbs. (= 1,937-lbs. per sq. in.) cracked in centre, kept at 50,000-lbs. (= 3228·3-lbs. per sq. in.) without crushing.
2	2·24	3·50	4·46	15·610	Layer of Sand.	Sign of a crack at 50,000-lbs. (= 3203 per sq. in.), kept at 52,000 (= 3362 per sq. in.) for 3 minutes and did not crush, and crack did not extend through brick.
3	2·34	3·50	4·52	15·820	Packed with Sand.	Crushed to pieces at 43,500-lbs. (2748·7 per sq. in).
4	2·34	3·46	4·46	15·4316	Packed with two pieces of cigar-box wood.	Edges crushed off at 30,000-lbs. (1990·1 per sq. in.)
5	2·30	3·46	4·50	15·570	Packed with Sand.	Cracked at 27,000-lbs. (1734·1 per sq. in.) Crushed at 32,000-lbs. (2055·3 per sq. in.), crushed and cracked in all directions, did not fall to pieces as did No. 3.
6	2·28	3·46	4·66	15·916	Packed with Sand.	Commenced to crack at 30,000-lbs. (1884·9 per sq. in.) Crushed to pieces at 46,500-lbs. (2921·6 per sq. in.)

FIRE-BRICKS.

Fire-Bricks should, on fracture, present a compact, uniform
structure—not necessarily close, for indeed some maintain that
a coarse grit of texture is the chief requisite, and that a close,
uniform structure, though pleasing to the eye, is not favorable
to the refractory powers of a fire brick, since the particles
should have a facility for contraction or expansion under heat,
and the air cavities act as valuable non-conductors of heat
The bricks should be free from stones, cracks, and irregular air-
hollows; and on being struck should emit a clear ringing
sound. The existence of this property usually involves a facility
for cutting.

Experiments made in the Royal Arsenal upon cubes of 1¼″ sides, cut from fire-brick "soaps" and placed between small squares of sheet lead, gave the following results:—

Description.	Cracked at.	Crushed at.
Stourbridge	1,478-lbs. per sq. in.	2,400-lbs. per sq. in.
Ditto	1,156 ,,	1,156 ,,
Newcastle	889 ,,	1,512 ,,
Plympton	1,689 ,,	2,666 ,,
Dinas	1,123 ,,	1,288 ,,
Kilmarnock	2,134 ,,	3,378 ,,
Glenboig	1,067 ,,	1,556 ,,

STOURBRIDGE FIRE CLAY.—Authority, S. Clegg, Jun.

The celebrated Stourbridge clay lies about 15′ beneath the lowest of three workable seams of coal (each averaging 6 feet thick) worked at Stourbridge in the lower coal measures in the south-western extremity of the Dudley coal field. The bed of clay is 4′ thick and the following is the composition:—

$$
\begin{array}{lr}
\text{Silica} . . . . & 63\cdot7 \\
\text{Alumina} . . . . . . & 22\cdot7 \\
\text{Oxide of Iron} . . . . & 2\cdot0 \\
\text{Water} . . . . . . & 11\cdot6 \\
\hline
& 100\cdot0
\end{array}
$$

(See also tests of bricks, Edinburgh Gas Works chimney, page 47.)

MORTAR.

EXPERIMENTS MADE BY GRAHAM SMITH, 1874.

"*Mason's mortar*" consisted of,—

1 slacked lime.   2 sand.   ¼ smithy ashes.

Being the ordinary mortar used in the construction of rubble masonry for dock walls.

"*Bricklayer's mortar*" consisted of,—

1 slacked lime.   1 sand.   1 smithy ashes.

The mode of testing pursued was as follows :—Bricks, the quality of which is described in each individual case, were accurately cut to $4\frac{1}{4}''$ in width, these were in all cases thoroughly wetted, and bedded crossways, with a mortar joint $\frac{5}{16}''$ thick and $4\frac{1}{4}'' \times 4\frac{1}{4}''$, giving a testing area of 18 sq. inches. On the time arriving for testing, which, unless otherwise mentioned, was 168 days, stirrups were passed round the ends of the bricks, two of these were attached to a beam and on the remaining two ends was hung a bucket into which perfectly dry sand was allowed to run from a hopper, the door of which was immediately closed when the joint parted. The bucket and sand were then weighed and this was taken to be the breaking weight of the specimen. In order to ascertain the difference which would exist in practice from the employment of bricks of various texture, two qualities were experimented on, viz., "common bricks," similar to, although slightly harder, than those known about London as "ordinary stocks," and "fire-bricks," very hard and much the same as Staffordshire blue bricks.

Description of		Breaking tensile strain per 4½-in. × 4½-in. section.
Bricks.	Mortar.	
Common	Mason's	496-lbs.
Fire-bricks	Do.	433-lbs.
Common	Bricklayer's.	610-lbs.
Fire-bricks	Do.	516-lbs.

The above are the average results of three experiments in each instance, from which it would appear that soft porous bricks are preferable for work subjected to a tensile strain.

A second series of tests were made by first subjecting the samples, twenty-four hours after being bedded, to a pressure of 56-lbs., and following this up with an additional 56-lbs. every day until 4-cwt. had been placed upon each. These broke as follows :—

Description of		Breaking tensile strain per 4½-in. × 4½-in. section.
Bricks.	Mortar.	
Common	Mason's	683-lbs.
Fire-bricks	Do.	403-lbs.
Common	Bricklayer's.	372-lbs.
Fire-bricks	Do.	423-lbs.

It was thought that the mortar would bear a greater strain after being compressed as in practice, but this was only borne out in the first case, the remaining three cases being considerably below the respective averages given before. It is feared that in placing on the weights the mortar was disturbed after having partially set, in which case it will never bind together a second time.

Samples were tested joined together with mortar re-mixed with water six days after the first mixing, it was found that,—

"Common" bricks with	"Mason's"	mortar re-mixed	. . .	broke at	432-lbs.
Do.	,,	do.	when first mixed .	,,	496-lbs.
Do.	,,	"Bricklayer's"	mortar re-mixed . . .	,,	440-lbs.
Do.	,,	do.	when first mixed .	,,	610-lbs.

The advantage is thus shewn of using mortar when first mixed.

The importance of the admixture of ashes with mortar, to be atmospherically dried, is shewn in the following tests:—

Description of		Breaking tensile strain per 4½-in. × 4½-in. section.
Bricks.	Mortar.	
Common	Bricklayer's with ashes .	570-lbs.
Do.	Do. no ashes .	257-lbs.

These two last tests were tried after a lapse of 84 days.

STONE.

The five following tables show the results of experiments to ascertain the resistance to *thrusting stress* of stones made for Mr. Samuel Trickett, of Millwall.

Laminated stones, as will be seen from the tests, are of far greater crushing pressure than homogeneous or rock stone; this was the result that Mr. James Gowan arrived at in his experiments given under head of Edinburgh Gas Works chimney, page 46, and in which Mr. S. Trickett agrees.

Description.	Dimensions.		Base area.	Cracked slightly.			Crushed, steel-yard dropped.		
				Stress.	Per sq. inch.	Per sq. foot.	Stress.	Per sq. inch.	Per sq. foot.
		Inches.	Sq in.	lbs.	lbs.	Tons.	lbs.	lbs.	Tons.
Leigh Carr, Lancashire. A laminated stone, the crushing test being equal to many Granites.	6·08	6·10 × 6·16	37·57	589,000	15,677	1008·1	617,000	16,422	1056·0
	6·08	6·06 × 6·10	36·96	572,000	15,476	995·2	594,360	16,081	1034·1
	6·06	6·10 × 6·10	37·21	564,000	15,157	974·7	593,870	15,959	1026·3
				575,000	15,437	992·7	601,743	16,154	1038·8
Wild Carr. A laminated stone.	6·10	6·00 × 6·00	36·00	376,300	10,452	672·1	415,360	11,537	741·9
	6 00	5·95 × 6·00	35·70	342,400	9,591	616·7	404,580	11,332	728·7
	6·05	6·03 × 6·03	36·36	354,500	9,749	626·9	409,750	11,269	724·6
				357,733	9,930	638·5	409,896	11,379	731·7
Appleton Quarries Shepley. A laminated stone.	6·00	5·95 × 5·98	35·58	341,600	9,600	617·3	372,890	10,480	673·9
	5·98	5·98 × 5·98	35·76	369,000	10,318	663·5	411,470	11,506	739·9
	5·95	5·94 × 5·95	35·34	324,500	9,182	590·4	350,850	9,927	638·3
				345,033	9,700	623·7	378,403	10,637	684·0
Corsehill Quarry, Annan, Dumfriesshire. A red laminated stone.	5·90	5·90 × 5·92	34·92	298,620	8,551	549·9	349,180	9,999	643·0
	5·98	5·99 × 5·98	35·82	263,180	7,347	472·4	285,840	7,979	513·1
	6·00	5·98 × 5·98	35·76	200,800	5,615	361·0	207,310	5,797	372·7
				254,200	7,171	461·1	280,776	7,925	509·6
Bramley Fall, Weetwood Quarries, Leeds. A coarse grit stone.	5·88	5·90 × 5·98	35·28	131,800	3,735	240·1	154,290	4,373	281·2
	6·16	5·98 × 6·02	36·00	126,930	3,525	226·6	148,370	4,121	265·0
	5·90	5·92 × 5·98	34·81	118,950	3,417	219·7	135,920	3,904	251·0
				125,893	3,559	228·8	146,193	4,132	265·7

All bedded between pieces of pine ¾″ thick.

Description.	Dimensions.		Base area.	Cracked slightly.			Crushed, steel-yard dropped.		
				Stress.	Per sq. inch.	Per sq. foot.	Stress.	Per sq inch.	Per sq. foot.
	Inches.		Sq. in.	lbs.	lbs.	Tons.	lbs.	lbs.	Tons.
Ancaster. Tested on the bed	12·00	12·03 × 12·00	144·36	235,300	1,629	104·7	310,788	2,152	138·7
	12·00	12·00 × 11·96	143·52	226,400	1,577	101·4	307,110	2,139	137·5
	12·00	12·00 × 11·93	143·16	174,800	1,221	78·5	208,570	1,457	93·6
			Mean.	212,166	1,476	94·8	275,486	1,916	123·2

All bedded between pieces of pine ¾″ thick.

(See also tests of stone, Edinburgh Gas Works chimney, page 46.)

PORTLAND CEMENT.

Weight and Sieve Test.—This cement should weigh at least 110-lbs. per imperial striked bushel, or 85-lbs. per cubic foot, filled from an inclined plane at an angle of say 45°. The weight test is also applied by filling the cement in through a hopper having a rectangular distributing shoot, the height or drop from the bottom of the hopper to the top of the bushel measure being 18″. Fine grinding, however, makes a most important alteration in the weight test, as the following table indicates.

An imperial bushel of best cement, freshly ground, passing through a,—

80 mesh sieve and leaving 10% residue, weighs 110-lbs.					
Do.	do.	20	„	„	116-lbs.
Do.	do.	25	„	„	121-lbs.
Do.	do.	35	„	„	123-lbs.

Cement and Sand Test.—Some engineers have lately discarded the weight test and rely upon test bricks made up as concrete, in the proportion of three of standard sand and one of cement, to be broken at 28 days after gauging, during one of which it has been in the air and 27 days in water. Samples thus mixed should have a minimum strength of 112-lbs. per square inch.

Samples made one of cement, weighing 123-lbs. per bushel, and one of Thames sand,—

1 week old,	broke at	160-lbs. per sq. inch.	
1 month „	„	201-lbs.	„
3 „ „	„	244-lbs.	„
6 „ „	„	284-lbs.	„
9 „ „	„	307-lbs.	„
12 „ „	„	318-lbs.	„
24 „ „	„	351-lbs.	„

Tensile Test.—When mixed up neat and immersed in water Portland cement should, after seven days, be capable of resisting a tensile strain of at least 600-lbs. per 2¼ square inches section. In making the test briquettes the proportion of water should be 9-oz. to 40-oz. of cement, both being accurately weighed, as an excess of water will adversely affect the result.

Chemical Analysis,—

Lime	59·13
Silica	23·67
Alumina	8·14
Per oxide of iron	3·26
Sulphuric acid	1·44
Alkali	2·46
Moisture, &c.	1·90
	100·00

A higher proportion of lime gives greater tensile strength at short dates, but 'is found not to be depended upon, although sometimes a long period may elapse before any deterioration becomes apparent.

Compression Test,—

3 months old . . .	1·71 tons per sq. inch.	
6 „ „ . . .	2·43 „	„
9 „ „ . . .	3·16 „	„

Mr. Faija, who is a well-known authority on Portland cement, has made a great number of crushing tests, and considers they are practically failures. There is a difficulty in the manufacture of the cube. If it is not perfectly true the weight first comes on one corner and then on another, and in fact the cube is attacked in detail, and the result of the test is therefore not of much value. Tensile strain bears a fixed proportion to other strains, and as the tensile strain is the easiest to apply, it is the most useful. Portland cement a week old will carry about 6,000-lbs. crushing strain, but in half-a-dozen tests one will get a variation often of 100 per cent.

Samples of best heavy Portland cement tested by Adie's machine, 1880; mould $1\frac{1}{2}'' \times 1\frac{1}{2}'' = 2\frac{1}{4}$ sq. inches.

Number.	Days after moulding.	Days in water.	Average tensile strain.
22 samples . . .	3	2	850-lbs.
22 ,, . . .	4	3	890-lbs.
42 ,, . . .	7	6	910-lbs.
42 ,, . . .	15	14	1,040-lbs.
36 ,, . . .	31	30	1,190-lbs.
22* ,, . . .	91	90	1,360-lbs.

* N.B.—Of these seven bricks unbroken at 1,440-lbs.

Samples of Portland cement tested by Henry Faija; June, 1884, mould $1'' \times 1'' = 1$ sq. inch :—

AVERAGE OF 25 TESTS.

	3 day test.	7 day test.
Tensile strain per sq. inch	362-lbs. . . .	404-lbs.
Equal to a strain on $1\frac{1}{4}'' \times 1\frac{1}{4}''$ of	814-lbs. . . .	909-lbs.

Samples of Portland cement tested by D. Kirkaldy & Son, June, 1884; mould $2\frac{1}{2}'' \times 2'' = 5$ sq. inches :—

AVERAGE OF 10 TESTS.

Tensile strain per 5 sq. inches . . . . 2118 to 2144-lbs.
Equal to a strain per $1\frac{1}{2}'' \times 1\frac{1}{4}''$ of . . . . 953·1 to 964·8-lbs.
,, ,, ,, ,, square inch . . . . . . 423·6 to 428·8-lbs.
Weight per imperial bushel . . . . 116·8-lbs.

Portland cement may be had guaranteed by the manufacturers to weigh and break as follows :—

Description.	Weight per bushel.	Minimum breaking strain, 1¼ in. × 1¼-in. section.
Best heavy Portland cement (slow setting) . .	116 to 120-lbs.	800 to 1,000-lbs.
Best quick setting cement	112-lbs.	600 to 800-lbs.

CONCRETE.

Tests of 6 inch cubes of Lias lime concrete, 53 weeks after being moulded,—

Lime = 71-lbs. and gravel 137½-lbs. per bushel.

C

The following are the average results of 10 tests in each case:—

Proportion.	Weight per cubic foot in lbs.	Crushed at Tons.	Tons per super ft.
6 to 1	131	5·77	= 23·08
8 to 1	129·6	2·69	= 10·76
10 to 1	128	2·14	= 8·56

DRAUGHT.

In Fig. 1 let L B = C R = the height of the atmosphere, F B = the height of a chimney, and A R = a similar column of external air equal in height to the shaft. The columns C A and L F of the atmosphere above the chimney are equal, and the pressures caused at A and F thereby may be disregarded. When the temperature and density of the air inside the chimney are equal to those of the similar column A R there will be no tendency to produce motion, should however the air inside the chimney be heated, and consequently become of less density than the column A R, there will be a greater pressure at G M by A R than by F B, and the excess of pressure will force the heated air in the shaft upwards. If the outside air as it enters the chimney at G M be heated and the diminished density of the air within the shaft maintained, the column of air F B will continue to ascend in consequence of the preponderating pressure of the outside column A R. This is what takes place when a fire is lighted at the base of a chimney and the unbalanced pressure of the external air causes what is known as the "draught."

The co-efficient of expansion of gases on the Fahrenheit scale is found to be $\frac{1}{491}$, or in other words by raising the temperature from freezing point to 491° above freezing point, the volume will be doubled and its density reduced by one-half. The volumes of the cold air and heated gases may be found from the following formula, and also the respective densities, since the densities vary inversely as the volumes :—

Let t = temperature of external air, say 60° F.
T = temperature of heated air in shaft.
d = density of　do.　do.
D = ·0765-lb., the density or weight of a cube foot of air at 60° F., with the barometer at 30″.

$$\text{Vol. at } T = \text{vol. at } t \frac{491 + T - 32}{491 + t - 32} = \text{vol. at } t \frac{459 + T}{519} \quad . \quad (1)$$

$$d = D \frac{519}{459 + T} = ·0765 \frac{519}{459 + T} \quad . \quad . \quad . \quad . \quad . \quad (2)$$

If we take J B as representing the volume of external air that would, if raised to the temperature of the air in the shaft, fill the chimney, we then have A D as the head or motive column of external air producing the draught.

If h = the height of the chimney in feet,
H = the head of external air in feet graphically shewn by $A\ D$,
Then

$$H = \frac{hD - hd}{D} = h\frac{D-d}{D} \quad \ldots \ldots \ldots \ldots (3)$$

And by (2) we have

$$H = h\frac{T-t}{459+T}\cdot \quad \ldots \ldots \ldots \ldots \ldots (4)$$

It has been found in practice that the best results in the draught are obtained when the density of the external atmosphere is to the density of the gases in the shaft as 2 to 1, or when the temperature of the escaping gas is about 580° F.; above this there is no practical gain in the draught and a great waste of fuel takes place, as well as injury to the brickwork. It is also found that with a lower exhaustion than $\frac{1}{4}''$ water pressure it is diffcult to keep a good fire without constant stirring, which is wasteful and productive of smoke.

Taking therefore the temperature of the heated air at 580° we have,—

$$H = \frac{h}{2} \quad \ldots \ldots \ldots \ldots \ldots \ldots \ldots (5)$$

To this head of external air is due the velocity of the air entering at G M, and by the law of falling bodies this will be equal to that due to the height A D = H.

Let v = theoretical velocity of entering cold air per sec.
g = acceleration generated by gravity per sec. = 32-ft.
Then

$$v = \sqrt{2\,g\,H} = 8\sqrt{H}\cdot \quad \ldots \ldots \ldots \ldots (6)$$

As the velocity of the gases is proportional to their volumes, and the volume of air we have taken to be doubled upon entering the shaft, we have the theoretical velocity of the hot air in feet per second,—

$$V = 2\,v = 16\sqrt{H} \quad \ldots \ldots \ldots \ldots \ldots (7)$$

Syphon water gauges are commonly used to take the draught of chimneys, the exhaustion being given in inches of water.

A cube foot of water at 60° F. weighs 62·5-lbs., therefore the density of air is $\frac{1}{817}$th that of water.

If W = the theoretical draught of shaft in inches of water pressure,
Then

$$W = \frac{H \times 12}{817} \quad \ldots \ldots \ldots \ldots \ldots \ldots (8)$$

And

$$v = 66\sqrt{W} \quad \ldots \ldots \ldots \ldots \ldots \ldots (9)$$

From this data we have the following table of the theoretical draught powers of chimneys with the external air at 60° and the internal heated air at 580°:—

TABLE A.

Height of Chimney In feet.	Draught in Inches of water.	Theoretical velocity in feet per second.	
		Cold air entering.	Hot air at outlet.
50	·367	40·0	80·8
60	·440	43·8	87·6
70	·514	47·3	94·6
80	·587	50·6	101·2
90	·660	53·7	107·4
100	·734	56·6	113·1
120	·880	62·0	123·9
150	1·101	69·3	138·6
175	1·285	74·8	149·6
200	1·468	80·0	160·0
225	1·652	84·8	169·7
250	1·836	89·4	178·9
275	2·020	93·8	187·6
300	2·203	98·0	196·0

This theoretical draught is in practice diminished by the contraction at the entrance of the ash pit and at the bed of fuel and over the bridge, by the bends in the flues and the friction of the air or gases passing over the sooty surfaces in the flues and chimney. From observations and experience it may be taken that where there are not many restricted passages the velocity as a rule is that given by using a co-efficient of ·3, then we have from (7),—

$$V_r = 4·8 \sqrt{H}$$

Table B has been calculated upon this basis, taking 10-lbs. of coal per h.p. per hour, and 300 cube feet, or nearly 23-lbs. of air at 60°, per lb. of fuel. In passing through the fire the air is highly heated and expands to about $3\frac{1}{2}$ times its former volume, as it travels to the shaft its temperature is lowered to about 580° and it is thus reduced to twice its normal volume, or to 600 cube feet. We thus have 6,000 cube feet of hot air ascending the shaft per horse-power per hour.

It will be found upon comparing the table with the examples given hereafter that many agree with the powers in the table, while many do not. In fact, by referring to the examples, the only conclusion one can arrive at is that many shafts are out of proportion to the work they have to perform.

In building a shaft allowance of say 25°/$_{0}$ in the power of the chimney should be made for future additions. In the large shafts, however, where the horse-power is considerable, the ·3 co-efficient may safely be taken to cover contingencies and the powers in the table worked to. It has been found that many large shafts have worked considerably better as more boilers have been added, and we can readily understand this being so up to a certain limit, as the additional furnaces will help to maintain a higher temperature and the gases not so readily cool. It may be here remarked that it is next to impossible to determine the actual velocity of the gases in a given shaft, as the conditions affecting the draught are daily, and we may say hourly, varying, and much depends on the quality of the coal used and the care displayed in firing. The direction of the wind also influences the draught as well as the temperature and density of the atmosphere.

TABLE B.

HORSE-POWER OF CHIMNEY SHAFTS TO STEAM BOILERS, FLUES NOT MORE THAN 180′ LONG.

Size of top inside		50-ft.		70-ft.		90-ft.		120-ft.		175-ft.		250-ft.		300-ft.	
		Square.	Round.	Square.	Round.	Square.	Round.	Square.	Round.	Square.	Round.	Square.	Round.	Square.	Round.
ft.	in.														
2	0	57	45	68	53										
2	3	74	58	87	68										
2	6	90	71	107	84	121	95								
2	9			129	101	146	115								
3	0			154	121	174	137	201	158						
3	6					237	186	274	215						
4	0					310	243	358	281	432	339				
5	0							560	440	675	530	807	634		
6	0							806	633	972	763	1163	913		
7	0									1323	1039	1583	1243	1735	1363
8	0									1728	1357	2067	1623	2265	1779
9	0											2616	2054	2867	2252
10	0													3542	2782

STABILITY.

Chimney shafts are exposed to the lateral pressure of the wind, tending to overturn the structure. This pressure may be assumed to act horizontally, and be of uniform intensity at all heights above the ground, without any appreciable error. The inclination of the surface of the chimney to the vertical is usually so small that it may be disregarded in estimating the pressure of the wind against the shaft. The greatest intensity of wind pressure used to be taken by Rankine at 55-lb. per square foot against a flat surface directly opposed to it. Although anemometers have registered much greater pressures than this, even as high as 80-lb. per square foot, we have it on the authority of Messrs. Fowler and Baker that the records of anemometers, as at present obtained, are utterly misleading and valueless for all practical purposes, and a gauge was made in the presence of a Board of Trade Inspector to register 65-lbs. by the sudden application of a pressure not exceeding 20-lbs., the momentum of the index needle sufficing to cause the error. Mr. B. Baker, in his paper on the Forth Bridge, read before the British Association at Montreal, 1884, says, "Mr. Fowler and I are of opinion, therefore, as a result of our two years' further consideration, that the assumed pressure of 56-lb. per square foot (recommended to be allowed for by the Board of Trade Committee on Wind Pressure) is considerably in excess of anything likely to be realised."

The pressure of wind against a circular shaft may be taken as being equal to half the total pressure against a diametral section of that shaft. This result is obtained as follows :—

In Fig. 2 let $d\,c = p$, the force of the wind acting parallel to the diameter $b\,a$. Resolve this force into its component parts, acting at right angles to one another at the point c, one of them, $f\,c$, being a normal to the curve, we then have $f\,c$ as representing the force of the wind acting towards the centre of the shaft, and $f\,c = p$ cos. $\angle\,d\,c\,f$. Resolving this force $f\,c$ at the point g, so as to measure the effective force exerted in the direction $g\,a$ parallel to the wind, we have the effective pressure $P = p$ cos.^{2} $\angle\,d\,c\,f$. This angle $d\,c\,f$ ranges from $0°$ to $90°$, and taking a sufficient number of angles we obtain cos.^{2} $\angle\,d\,c\,f =$ about $\cdot 5$, therefore the mean effective pressure of wind against the semicircumference, $P = \cdot 5\,p$.

In this manner we obtain that if—

The pressure on a square shaft be taken $= 1$
That on a hexagonal shaft may be taken $= \cdot 75$
That on an octagonal shaft may be taken $= \cdot 7$
That on a circular shaft may be taken $= \cdot 5$

If it is required to determine the stability of that portion of a chimney shaft above the bed joint $c\,d$ (Fig. 3),—

Let $A =$ the area of the diametral section of the shaft above $c\,d$, then the pressure of the wind against the shaft will equal—

$$P = p\,A \text{ for a square chimney,}$$
$$P = \cdot 5\,p\,A \text{ for a round chimney,}$$

And its resultant may be taken as acting in a horizontal line through e, the centre of gravity of the diametral section. Let H represent the height of e above $c\,d$, then the overturning moment is—

$$P\,H = p\,A\,H \text{ for a square chimney,}$$
$$P\,H = \cdot 5\,p\,A\,H \text{ for a round chimney,}$$

And the *least* moment of stability of the shaft above $c\,d$ should be equal to this.

It is evident that this lateral pressure of the wind will tend to move the centre of pressure on the joint $c\,d$, towards the lee side. It is found, in practice, advisable so to limit the deviation of the centre of pressure from the centre of figure that the maximum intensity of pressure at the leeward side shall not exceed twice the mean intensity. Let q denote the ratio which the distance of this deviation bears to the length of the joint $c\,d$, then we have the following value as given by Rankine:—

For square chimneys $q = \frac{1}{3}$,
For round chimneys $q = \frac{1}{4}$,

Which is practically taking a factor of safety of 2 for round shafts and $\frac{3}{4}$ for square shafts.

If we take a chimney the axis of which is not vertical, as in Fig. 3, it is evident that the least moment of stability is that which resists the overturning action of the wind in the direction in which the shaft leans. Let g be the centre of gravity of the part of the shaft above $c\,d$, f a point in the joint $c\,d$, vertically below g, a the limit of deviation of the centre of pressure, j equal the length of $c\,d$, q^1 the ratio which the deviation of f from the middle of joint bears to j, W the weight of the shaft above $c\,d$; all the values being in feet and lbs.; then the *least* moment of stability is,—

$$W \times f\,a = W\,(q - q^1)\,j$$

Which should equal the moment of wind pressure. Therefore we have the equation,—

$$P\,H = W\,(q - q^1)\,j.$$

Substituting the values of $P\,H$ and q, this becomes,—

$p\,A\,H = W\,(\tfrac{1}{3} - q^1)\,j$ for square chimneys.

$\dfrac{p\,A\,H}{2} = W\,(\tfrac{1}{4} - q^1)\,j$ for round chimneys.

Let T be the mean thickness of the brickwork above the joint $c\,d$, and t the thickness to which the brickwork would be reduced if spread out upon a flat area equal to the external area of the shaft. This reduced thickness is given approximately by

$$t = T\left(\frac{j - T}{j}\right).$$

In most cases, however, the difference between T and t may be neglected.

If w be the weight of a cubic foot of brickwork $= 112$-lbs. generally, we have,—

$W = 4\,A\,t\,w$ for square shafts,

$W = 3\text{·}14\,A\,t\,w$ for round shafts,

And substituting these values in the equations above given, we obtain,—

$p\,H = (\tfrac{4}{3} - 4\,q^1)\,t\,w\,j$ for square chimneys.

$p\,H = (1\text{·}57 - 6\text{·}28\,q^1)\,t\,w\,j$ for round chimneys.

If we consider the chimney to stand vertically on its base, this becomes,—

$p\,H = \tfrac{4}{3}\,t\,w\,j$ for square shafts.

$p\,H = 1\text{·}57\,t\,w\,j$ for round shafts.

In the above formulæ the tenacity of the mortar has been disregarded, and it should never be taken into consideration in designing new shafts, as many months from erection must elapse before the tenacity of the mortar is appreciable.

The foregoing formulæ enables us to determine the greatest pressure a shaft will withstand when we have the dimensions, forms, and thicknesses of the masonry or brickwork of the chimney given, and also to find the value of t for each bed joint when we have the pressure of the wind p and the external form and dimensions of the chimney given.

A chimney shaft consists of a series of sections one above the other, each section being of uniform thickness, and each succeeding section diminishing in thickness from that immediately below it, and it is obvious that the bed joints dividing the sections have less stability than the intermediate ones; hence it is only necessary to apply the formulæ to the former set of joints, including the joint at the ground line.

The stability against wind of Messrs. Tennant and Co.'s chimney, St. Rollox, Glasgow, Prof. Rankine determined by the formulæ herein given.

DETAILED EXAMPLES OF EXISTING SHAFTS

TOWNSEND'S CHIMNEY, PORT DUNDAS, GLASGOW,

THE TALLEST CHIMNEY IN THE WORLD.

Figs. 4, 5, 6 *and* 7.

Designed and built by Mr. ROBERT CORBETT, Bellfield Terrace, Duke Street, Glasgow, for Mr. Joseph Townsend, Crawford Street, Chemical Works, Port Dundas.

Dimensions,—

Total height from bottom of foundation to top of coping .	468′ 0″
Height from ground line to top of coping	454′ 0″
Outside diameter at ground line	32′ 0″
,, ,, ,, top	13′ 4″

The shaft from ground line to top is divided into 12 sections, the thickness of walls varying from 5′ 7″ at bottom of shaft to 1′ 2″ at top, viz.:—

1st section,	30 feet in height	5′	7″ thick.
2nd ,,	30 ,,	,, ...	5′	2″ ,,
3rd ,,	30 ,,	,, ...	4′	10″ ,,
4th ,,	40 ,,	,, ...	4′	5″ ,,
5th ,,	40 ,,	,, ...	4′	0″ ,,
6th ,,	40 ,,	,, ...	3′	7″ ,,
7th ,,	40 ,,	,, ...	3′	2″ ,,
8th ,,	40 ,,	,, ...	2′	9″ ,,
9th ,,	40 ,,	,, ...	2′	4″ ,,
10th ,,	52 ,,	,, ...	1′	11″ ,,
11th ,,	52 ,,	,, ...	1′	7″ ,,
12th ,,	20 ,,	,, ...	1′	2″ ,,

Total . . 454 feet from ground line.

The height originally contemplated for the chimney was 450′, but when 350′ had been built, it was proposed to add about 35′ to the original height, making a total of 485′, hence the increased height of the 10th and 11th sections, but on completion of the eleventh section this idea was abandoned, and therefore only 20′ of the last thickness were added.

(Fig. 4.)—*Foundations.* No piles were used. The blue till or clay on which it is founded being as solid and compact as a rock. The circular foundation consists of 30 courses of brick on edge, giving a height of about 12′. The lowest course is 50′ diam. and the topmost 32′. Four eliptical ports were formed in the foundation each 6′ 0″ × 7′ 6″, brickwork to arch and invert being 14″ deep.

The calculation of Prof. Rankine was that each brick, from outside to centre, bore equal pressure.

The total weight of chimney and foundation, including mortar, bricks, hoop iron, &c., was estimated at 8,000 tons.

Bricks.—These were made of good clay ground with 10°/₀ good clinkers.

The chimney is built entirely of bricks, from foundation to cope; size 10″ × 4″ × 3½″, weighing 5 tons per 1,000 bricks.

Chimney	1,142,532 common bricks	= 5713	tons.
„	157,468 Scotch fire-bricks for inner cone	= 787	„
	1,300,000	6,500	„
Flues, &c.	100,000	500	„
Total bricks,	1,400,000	= 7,000	„ for bricks only.

The bricklayer's time, in building the chimney alone, was,—

In 1857 . . . . 316 days of 10 hours each.
1858 . . . . 431½ „   „   „
1859 . . . . 423½ „   „   „

Giving a total of 1,171 days, which was on an average 1,200 bricks laid per day of ten hours.

The bricks were tested before use, and withstood a far greater pressure than that estimated to come upon them.

Mortar,—

1 Good Irish lime, well burnt.
3 „ sharp sand.

Added to the above was a portion of fine red oxide of iron.

The mortar was of ordinary consistency. The bricks were wet before use, and each course when laid was grouted with lime. The joints were of medium thickness.

Hoop Iron Bond.—Iron hoops 4½″ × ¾″ were built edgeways in the brickwork every 25′, 9″ from the outer surface at the base and 4½″ from outer surface at the top.

Stability.—Prof. Rankine, in his report upon the stability of this chimney, calculated the maximum pressure of wind which this structure is capable to resist at different horizontal joints, and the figures given by him are as follows:—

Height of joint above ground.	Greatest safe pressure of wind.
360 feet	90-lbs. per sq. ft.
280 „	64 „ „
200 „	63 „ „
120 „	66 „ „
40 „	74 „ „

Building Operations,—

1st SEASON.

Foundations commenced 30th July, 1857.
,,   finished, 20th Aug., 1857.
Erection continued till 11th Nov., 1857,

Except from 3rd Sept. to 5th Oct., during which period operations were suspended.

2nd SEASON.

Resumed operations, 10th June, 1858.
Suspended operations, 16th Oct., 1858.

Chimney at this stage about 228′ in height.

3rd SEASON.

Resumed operations, 3rd June, 1859.
Finished operations, 6th Oct., 1859.

Chimney was raised 226 feet this season.

The building operations were suspended from the 15th Sep. to 5th Oct., in consequence of the chimney swaying. During this interval it was restored by No. 12 cuttings, with saws on the opposite side to inclination, as hereinafter mentioned under head of straightening.

Coping. — The cope was formed of Garnkirk fire clay, purposely made about 9″ wide by 3″ thick, flanged over wall of chimney (see diagram No. 7) and jointed in Portland cement.

The strength of this coping was such that in a storm some pieces were displaced by lightning and fell to the ground, and upon inspection it was found only one of the flanges was broken.

Inner Cone of fire-brick lining. An inside lining or cone of 9″ fire-brick, about 60′ in height, was built distinct from the chimney proper with a 9″ air space between, covered in on top to prevent dust falling in. The four upper courses of the lining were built in open work to allow of the free circulation of air.

Cost.—The chimney was constructed by day work, as Mr. Townsend was not certain to what height he would carry it, or what deviation would be necessary from the proposed plan. Three different sets of dimensions were roughly calculated by the builder; those carried out were the greatest of the three.

The builder reckons the cost of the chimney and cone was —exclusive of iron hoops and flues—about £5,500 to £6,000. Mr. Townsend estimates the cost of the whole, including flues, iron hoops, machinery and scaffolding (£700) was £8,000. To this latter amount the builder assents.

Flues, eliptical in form. They carry off a variety of gases, arising formerly from the burning of bones, old horses, &c.,

for manure, and also the ordinary smoke. Now, however, there are only the gases from ordinary chemical operations (say drysalteries) and smoke, the manure operations having been given up.

The flues branch to all parts of the works, which cover an area of nearly 4 acres.

Straightening.—On Sept. 9th, 1859, the chimney (height 449′) was struck by a gale from the N.E., which caused it to sway. The builder does not attribute this action to the gale alone, but to the pressure of the whole pile on the scaffolding, which was so constructed as not to yield to any pressure caused by settling down. The additional pressure caused by the wind on the lee side of the stalk (the mortar of which was not set) was consequently too great for the scaffolding to bear, and caused the splices of one of the uprights—A diagram, No. 6— to give way, making the fibres of the timbers to work into each other by compression. The ends of the bearers—B diag. No. 5— were tightly built into the masonry at each staging, which occurred every 5′ to 6′; had a space of say 4″ been left over each end of the bearers the stalk would have subsided uniformly, and would possibly have withstood the gale. The builder observed this omission, but too late; he thinks the deflection commenced at from 100′ to 150′ from the ground, so that the foundation and heavier portion remained firm. The chimney would have probably fallen had not the process of sawing been commenced promptly and continued vigorously. Even during the earlier part of the process of sawing, Mr. Townsend, who was on the ground the whole time, observed the deflection increasing, but as the sawing progressed he noticed it received a check and the shaft "came to" gradually. The chimney was bent 7′ 9″ at top from its original position, and was less in height than before it swayed, but when brought back regained its former altitude.

Mr. Townsend made his observations during the sawing-back by taking a position in a room of the works near the chimney where he had a full view of it, and fixing the ends of two pieces of twine to a beam above, he formed them into two plummet-lines in a line with the stalk, and with these alone he directed the adjustment of the colossal mass.

The sawing-back was performed by Mr. Townsend's own men from the inside, on the original scaffolding, which, of course, had not been removed. Holes were first punched through the sides to admit the saws, which were worked in opposite lateral directions from such holes at the same joint.

This was done at twelve different heights from the ground line, viz.: 41—81—121—151—171—189—209—228—240—255—

277′ and 326′. The chimney was brought back in a slightly oscillating manner, and the men discovered when they were gaining by the saws being tightened by the superincumbent weight. It took six men continuously working at the sawing-back—four sawing and two watering—at a total cost of £400.

Prior to the sawing operations the bolts of the scaffolding were taken out and altered, so as to relieve the pressure on it. This was done to meet the want of a little space over the ends of the uprights, as before stated.

Lightning Conductors.—There are two ⅞″ copper lightning conductors attached to a copper spire 6′ to 8′ above the top of chimney. Round the top of chimney there runs a copper coil connecting the conductors.

At the bottom the two conductors are led to a well, which is kept moist, and a bar of iron 3″ square and 8′ long is driven into the earth and connected with the conductors.

Completion.—For many days after the chimney was brought back to the perpendicular and finished Mr. Townsend invited the public to go to the top, and thousands availed themselves of the opportunity. Parties of two at a time were sent up on a small platform without sides, and having at a convenient height a circular cross-bar, on each side of which one person stood and held on. It was quite dark from the time of leaving the ground until emerging through the hatch at the top. There were always four at the top and two going up; when they arrived two came down to make room. The machinery used for hoisting visitors and materials was driven by friction gearing, an ordinary strong rope being employed.

Between 200 and 300 persons were sometimes waiting at the base of the shaft, so great was the rage to mount the monster "lum," and many waited half a day before ascending.

MESSRS. TENNANT & CO.'S CHIMNEY, ST. ROLLOX, GLASGOW.

Figs. 8 *and* 9.

The chimney shaft of Messrs. Charles Tennant & Co., St. Rollox Chemical Works, Glasgow, was projected by Professor W. J. MACQUORN RANKINE; designed by Messrs. L. D. B. GORDON and L. HILL, and built by Mr. McINTYRE.

Dimensions,—

Total height from bottom of foundation to top of coping	455½′
Height from ground line to top of coping	435½′
Outside diameter at foundation	50′
,, ,, ground line	40′
,, ,, at top	13½

The height from ground line to top is divided into 5 sections, the thickness of walls varying from 2′ 7¼″ at bottom of shaft to 1′ 2″ at top, viz.:—

Dimensions and stability of outer shell, as given by the projector, Prof. RANKINE:—

Divisions of Chimney.	Height above ground line. ft.	External diam. ft. ins.	Thicknesses. ft. ins.	Greatest pressure of wind consistent with security. lbs. per sq. foot.
5 ...	{ 435½ ... 13 6 }		1 2	
4 ...	{ 350½ ... 16 9 }		1 6	77
3 ...	{ 210½ ... 24 0 }			55*
2 ...	{ 114½ ... 30 6 }		1 10½	57
1 ...	{ 54½ ... 35 0 }		2 3	63
	{ 0 ... 40 0 }		2 7½	71

Foundation.	Depth below ground. feet.	External diam. feet.	Thicknesses. Concrete. ft. ins.	Bricks. ft.
1 ...	{ 0 50		5 0	3
	8 50		4 8	3
2 ...	{ 14 .... 50 }		25 0	8
3 ...	{ 20 .... 50 }			0

* Joint of least stability.

Foundation.—No piles were used, it being founded on free-stone. The construction of circular foundation of outer cone and the square foundation for inner cone, will be best understood by reference to the diagrams, Figs. 8 and 9, together with Professor Rankine's dimensions given above.

The shaft was founded on the site of an old quarry, and the ground on the north side had to be concreted up to level of rock foundation.

Pressure at the base of chimney, 450′ below summit.

	Pressure in lbs. on the sq. ft.	on the sq. in.
On a layer of strong concrete or béton, 6′ deep	6,670	46
On sandstone below the béton, so soft that it crumbled in the hand	4,000	28

The last example shows the pressure which is safely borne in practice by one of the weakest substances to which the name of *rock* can be applied.

Bricks.—The chimney is built entirely of bricks excepting coping, which is stone.

There were used 1,250,000 bricks, weighing 121-lbs. per cubic foot, and resisting 63 tons pressure per super foot before cracking. The bricks were of the first quality, selected and purposely made of a mixture of ironstone and blue clay.

Bond.—The bond used in its construction is "old English."

Inner Cone.—The inner cone is built up of 3 sections, with best bricks, same as main shaft (iron-stone and clay composition) of the following dimensions :—

1st section	60' 22¼" thick.	
2nd ,,	80' 18"	,,
3rd ,,	103' 14"	,,

243' total from ground line.

The use of the inner cone was intended to protect the principal stalk at base from the heat and effects of the gases at their entrance into the shaft, but it is now considered to be a mistake.

Height of inner cone from foundation to top.	263'
,, ,, ,, ground line ,,	243'
Inside diameter at foundation.	12'
,, ,, top	13½'

Mortar.—The mortar was made with Arden lime and good sharp river sand,

Building Operations,—

Chimney founded 29th June, 1841.
Cope laid 29th June, 1842.

A roof was put over the work during the winter months, but was blown down during a gale. The shaft in consequence had to remain uncovered until building operations were resumed.

Scaffold.—Inside scaffolding was used.

Flues are circular and of various lengths, branches extending to all parts of the works, which cover an area of 80 acres.

Iron Hooping.—In May, 1844, a rent was discovered in one side about 36' long, extending from a point about 100' from the top downwards. This rent is said to have been caused by the heat and gases passing upward between the inner and outer shafts owing to one of the flue connections having become cracked. The rent gradually increased during June and July, and then a similar rent was observed on the other side, beginning somewhat lower down than that first discovered, but extending only 45'. This created some apprehension, and in August it was determined to examine the chimney where the cracks had occurred. Scaffolding appeared at first to be the only means of effecting the examination without stopping the works. Balloons were proposed, but the idea was rejected, though the late Mr.

Green offered his personal superintendence of the ascent. Scaffolding would have cost £20. Mr. Colthurst, C.E., suggested that by driving staples into the joints of the brickwork a man might be able to climb to the top safely and very cheaply. A man was obtained who carried out the suggestion, and actually went up on the outside of a neighbouring chimney 112′ high and repaired the coping ; the ascent and work occupying two days. Working upon this suggestion, a climbing machine was devised by Mr. L. Hill for examining the rent at a distance of 280′ from the ground. On gaining the position of the rent a strong pulley was fixed in the chimney, through which a rope was passed extending to the ground, so that to this point an ascent could be made at any future time. The persons employed were slaters, an old and a young man. It was made a principle not to bribe anyone to undertake the job. The men worked for 5s. a day, or little more than their ordinary wages.

Though the fissure was found in one place to be 2″ wide, yet the nature of it was such that a rod could not be put through the crack to the inside of the chimney. It would have been very desirable to have inserted a thermometer into the interior, but as the instrument would not pass through the rent it was not thought advisable to drive a hole for the purpose. It may be mentioned, however, that red hot matter has more than once been projected through the top of this high shaft.

Thirteen iron hoops were put round the shaft, in consequence of the cracks, by means of the climbing machine above mentioned, the chimney continuing in work the whole time. Additional hoops, making a total of thirty, had to be put round the shaft about 1871, when the chimney was struck by lightning. These additional hoops were put on by Mr. J. Wright, of Aberdeen, who had to fly a kite over the top to commence operations.

Lightning Conductor.—Fixed 1872, by Mr. R. Hall. The conductor consists of two glass insulated copper ropes, one on each side of the shaft, connected at the base to a 1¼″ square iron rod, carried underground through the works till it reaches the canal.

MESSRS. DOBSON & BARLOW, KAY STREET MACHINE WORKS,
BOLTON, LANCASHIRE.

The large chimney stack here is connected with seven boilers, ventilating flues, furnaces, &c. It was completed in Nov., 1842, and was then intended to serve a chemical works. Shortly after, the ground occupied by the chemical works was added to Messrs. Dobson & Barlow's establishment, and consequently the chimney came into their possession.

The following are some principal particulars:—

Description.—Octagonal, brick.

Dimensions,—

Total height from ground level	367' 6"
Girth at bottom	112' 0"
Each side of octagon at bottom	14' 0"
Girth at top	44' 0"
Each side of octagon at top	5' 6"
Thickness of brickwork at bottom	8' 0"
,, ,, top	1' 6"

Materials.—800,000 bricks and 120 tons of stone work were used in the building. The top cornices and mouldings required 30 tons of stone and cement.

Cost.—£3,000.

MESSRS. WESENFIELD & Co.'S CHIMNEY, CHEMICAL FACTORY, BARMEN, PRUSSIA.

Fig. 10.

Description.—Square brick pedestal, octagonal brick shaft.

Dimensions,—

Total height from foundation to top	345' 0"
Height from ground line to top	331' 0"
Pedestal 20' sq. × 40' high × 7 bricks	5' 3" thick.
Octagonal shaft 291' high.	
,, ,, 17' outside diam. at base × 5 bricks	3' 9" ,,
,, ,, 11' ,, ,, at top × 2 ,,	1' 6" ,,
Diminution of shaft 2½" in diam. every 10'.	
Batter 1 in 97.	
Internal octagonal clearance 8' throughout.	

Foundation.—This is on a bed of hard and coarse gravel, and made of large flat quarry stones bedded with "terrass" mortar in the proportions of 1 lime, 1 river sand, and 1 "terrass" (a description of pozzuolana).

Pressure on lowest part of chimney proper = 21,335-lbs. or 9½ tons per square foot.

Pedestal and Shaft.—Built with bricks and ordinary mortar, 1 lime to 2 river sand, prepared every morning by the masons.

D

On rainy days cement mortar was used in the proportion of 1 cement to 2 river sand. The courses of brickwork were flushed up with cement as construction proceeded. The crown of shaft was built with cement exclusively.

Building.—The foundation and pedestal were built in the summer of 1867, and the construction of the chimney was successfully completed in October of same year.

According to original design it was intended to build to a height of 260', but as the erection was proceeding in a very satisfactory manner it was considered safe to increase the height without altering the dimensions of the base, but before doing so, a comparison was made between the pressure on the foundation of this chimney and the pressure on the foundation of chimney erected at Bochum, Prussia. It was found to be as follows :—

	Lowest part of chimney proper.	
	Press. pr. sq. ft.	Press. pr. sq. in.
Chimney at Barmen, Prussia	21,335-lbs.	149-lbs.
„ at Bochum, „	18,429-lbs.	128-lbs.
Excess of pressure on Barmen chimney foundation	2,906-lbs.	21-lbs.

The three masons who constructed the chimney daily changed their positions, so as to equalise any unevenness in their respective laying.

Every fifty feet a course of brickwork was painted black, so as to indicate the height of any point of the chimney above ground.

The chimney was built from the inside. The materials were hoisted by a steam engine erected temporarily near the place of construction. The frame which supported the upper drum over which the chain worked was moved higher after the completion of every three or four courses, and was at the same time turned horizontally from one side of the octagon to the next one, so as to equalise the pressure of the frame on the masonry. The holes made into the masonry to support the frame were filled up with bricks and mortar immediately after the removal of the frame to a higher level.

The chimney when completed (Oct., 1867) was vertical.

In the spring of 1868, remarkable for storms and long-continued gales, this stalk inclined toward the north-east. The action of the south-west wind was probably aided by the softness of the mortar and the large size and shape of the ornamented chimney crown, which caught the wind and acted as on a long lever.

The deflection was considerable at the end of May and apparently increased.

As before mentioned, layers of bricks in shaft at distances of fifty feet from each other were painted black. The height of these black lines above the pedestal being known, they were, by means of a theodolite, projected on a board, which was fixed on the pedestal, and these projections showed that at,—

251' high it was out of plumb			45"
210' ,,	,,	,, ...	30"
160' ,,	,,	,, ...	16"
110' ,,	,,	,, ...	5"

The pedestal stood perpendicular. As the canting of the shaft was still increasing, immediate action had to be taken.

The ordinary method of straightening chimneys was at first resorted to.

A hole was made through the whole thickness of the masonry on side of chimney which required lowering, at a distance of four feet above the top of pedestal; into this hole a saw was passed, and an attempt was made to cut through one half of the shaft, but owing to thickness of wall and hardness of the bricks the saw could only be worked from one end, and the effect of sawing, after two hours' work, was almost *nil*.

The hole through the stalk having been made with little trouble, and the difficulty experienced in sawing, led to the idea of removing a course of bricks and, re-placing it by a thinner one. Before the work was proceeded with an experiment was made on an old inclined shaft, 120' high. This proving successful, it was determined to treat the new chimney in the same way.

Straightening.—A layer of bricks was broken out by means of pointed cast steel bars varying from 1½' to 5' in length.

Fig. 10 shows a horizontal section of this layer, the numbers 1, 2, 3, 4, &c., indicating the order in which the brickwork was removed.

When division 1 was broken out it was replaced by thinner bricks covered with "terrass" mortar. After this the two divisions marked 2 were broken out and replaced by thinner bricks; then the two divisions marked 3, and so on until one-half of the whole course had been exchanged.

Purposely-made flat shovels with long handles were used to lay the bricks, which had to be placed near the inside of the chimney. A side space of 5" was left between the newly-laid bricks and the old ones of the next division, so as to enable the workmen to break out the latter with greater facility.

The width of each single division was 2' to 2¼'. The masonry directly above was sufficiently dry not to give way when a course of that width was removed from below it. The replaced bricks were thicker near the points A and C, so that the difference was greater in the middle, and gradually less toward the extremities A and C.

As soon as the slit reàched these points the chimney began to move, and by slight oscillations slowly settled down on the new layer of bricks.

The time occupied in settling by oscillation at each substituted course varied from 18 to 36 hours, according to the widths of the slits, which were different in the various cuts performed.

The oscillations were the greater the higher the cut.

At the highest cut, 100-ft. from the top, the oscillations frightened the masons, and they left the work. The slit became alternately wider and narrower by three-quarters of an inch. This seemed to prove the elasticity of the structure.

The four cuts made were as follows:—

```
1st  . . . .  4' above pedestal greatest width  . . . .  ¾
2nd  . . . . 100'  „      „      „      „      . . . . 1¼
3rd  . . . . 140'  „      „      „      „      . . . . 1½
4th  . . . . 191'  „      „      „      „      . . . . 1¾
```

After the completion of this work the chimney continued during several weeks to settle slightly in the direction opposite to its former inclination. This circumstance had to be carefully considered beforehand, or else the slits would have been made too wide, and have produced an inclination in the opposite direction.

A severe storm on the 6th and 7th Dec., 1868, which overthrew several chimneys in the neighbourhood, did not affect this one.

The result of the straightening operation described above was quite satisfactory.

The heights of the upper cuts were reached as follows: Standing on a platform the masons made a number of holes into the exterior wall of the chimney 4' above the platform on which they stood. Into these holes the ends of iron bars were fixed, and boards secured to them so as to form another platform. Standing then on the latter they fixed another platform 4' higher in the same way. Every second platform was removed, so that the remaining ones were 8' apart; they were then joined by ladders for the workmen to ascend.

This method of straightening is only practicable when the chimney has a considerable diameter, and when the mortar is sufficiently dry not to give way under pressure of the bars and platforms.

In Dec., 1868, a chimney was straightened at Duisburg by the method just described, but as the diameter was not so great as that at Barmen, and as the mortar was soft, a wooden scaffold was erected round the shaft to reach the upper points, which required cutting. The breaking out and replacing bricks could not be done there in divisions wider than 5 to 10 inches, as the upper masonry not being dry would have settled down. When the chimney was straight a further settling down towards the side of the cuts was prevented by driving iron wedges, covered with mortar, into the slits.

EDINBURGH GAS WORKS, CHIMNEY.

Designed by Mr. MARK TAYLOR, Engineer to the Company. Mr. GEO. BUCHANNAN, C.E., and Prof. GORDON, of Glasgow, were consulted.

Builders,—

    Stonework . . Mr. JAMES GOWAN, of Edinburgh.
    Brickwork . . Mr. JAMES BOW, of Pollocksfields, nr. Glasgow.

Description.—Square stone pedestal, circular brick shaft.

Dimensions,—

Stone foundation under ground	6½'
Part of base ,, ,, ...	6'
Stone pedestal above ,, ...	65'
Brick shaft	264'
Total height from foundation to top	341½'
Height from ground line to top	329'

Foundation.—Stone.

             40½' sq. × 6½' deep.

The distributed pressure on the bottom of foundation per square foot = nearly 2½ tons.

Pedestal.—Stone,

            30' sq. at ground line.
            27¾' ,,   top.
            22½' internal diam. at bottom.
            20½   ,,   ,,   top.

This was built during the summer, at the end of which the works were suspended until the following year.

Brickwork.—Commenced and finished in the summer following the erection of pedestal.

OUTER BRICK SHAFT (circular).

Outside diam. at bottom . . .	26′	3″
Inter. „ „ . . .	20′	5″
Outside diam. at top . . .	15′	0″
Inter. „ „ . . .	12′	0″

This was built up in five steps as follows :—

1st section 35′ high	. . .	3½ b.	= 35″
2nd „ 40′ „	. . .	3	= 30″
3rd „ 48′ „	. . .	2½	= 25″
4th „ 58′ „	. . .	2	= 20″
5th „ 83′ „	. . .	1½	= 15″

264′ total.

The greatest pressure on any part of the work comes at the
lowest section, where it amounts to about 8 tons 2 cwt. per square
foot.

INNER BRICK SHAFT (circular).

This is distinct from the outer shaft and is 90′ high with 13′
internal diameter throughout, and was built in four steps, viz. :—

1st section 14′ high 35″ thick.
2nd „   6′ „ 30″ „
3rd „ 30′ „ 25″ „
4th „ 40′ „ 20″ „

90′ total.

The thicknesses include a lining of fire-brick 10″ thick for
20′, and 5″ thick for remaining 70′.

Materials.—The weight of, is about 3,700 tons.

Cost.—Total was £4,637.

Lightning Conductor.—Solid copper rod ¾″ diam.

Stone.—The stones used in the foundation are Cragleith,
Humbie and Hailes, which before use were tested by Mr.
Buchannan and Mr. James Gowan. The tests were made in a
most careful way by crushing cubes of 1″ square.

RESULTS OF TESTS.

Cragleith	crushed at 315 tons per sq. ft.
Humbie	„	240	„	„
Hailes	„	225	„	„

A second test of Cragleith showed that before being crushed
to powder it sustained a pressure of 440 tons per square foot.

The appearance, after fractures of the different cubes, was
that of a pyramid or wedge, and this led Mr. Gowan to assert

that if the cubes were enlarged a greater increase of strength would be gained, and further, that if the pressure were *vertical* to the line of cleavage a greater resistance would be obtained, so that such a stone as Hailes, which is a laminated stone, would increase in strength according to its surface more in proportion than that of a liver rock stone such as Cragleith. This led to discussion and further tests, the result being that with a 4″ cube from Hailes quarry the resistance was = to 567 tons per square foot.

Bricks.—These were supplied by Mr. Livingstone, of Portobello Brickworks, and were tested with the following results:—

Description of specimen.	Length. Ins.	Breadth. Ins.	Thickness. Ins.	Weight. lbs.	Crushing weight on ea. brick. tons.	Crushing weight per sq. ft. tons.
Extra size and quality	10	5	3	10¼	153	440
Do. do.	9½	4¾	2¾	9⅛	140	448

Repairs.—In May, 1874, a "Steeple Jack" was engaged to examine the state of shaft. Recourse was had to kite-flying to fix at top a rope by which the man could draw himself up, to the free end of which a couple of 56-lbs. weights were attached. It was found that the stack near top had been split in several places by lightning and iron hooping was necessary.

MESSRS. EDWD. BROOKS & SONS, FIRE CLAY WORKS, HUDDERSFIELD.

Engineer, ROBERT MORGAN ; *Builder,* JOHN STOCKS.

Description.—Octagonal brick panelled pedestal, with stone mouldings ; circular brick shaft.

Dimensions,—

Height from foundation to top . . . . 330′ 0″
„   „   ground line „ . . . . 315′ 0″
„   „   „   „ of pedestal . . . . . 70′ 6″
Outside measurement above plinth . . . . 27′ 0″
Inside   „   „ . . . . 15′ 0″
Outside diam. at top under cap . . . . 12′ 0″
Inside   „   „   „ . . . .   9′ 0″
Outside „ top of pedestal . . . . 20′ 4‵
Inside   „   „   „ . . . . 14′ 0″

Foundation,—

Concrete, 3′ deep, 36′ square.
Ragstone footings 36′ „
Brick   „   35′ „
Stepped to . . 31′ „   at ground line.

Materials,—

144 cu. yds.	.	concrete.
2452 ,, ft.	.	ragstone footings.
3341 ,, ,,	.	ashlar.
2227 ,, yds.	.	brickwork.

Inner Shaft.—An inner shaft is constructed for a height of 150′, divided at base with a wall 60′ high, to prevent baffling of the draughts. Round this inner shaft is a cavity which is supplied with cold air at the base.

Duty.—The chimney now serves two boilers, 17 kilns 22′ diam.

Cap.—The original cap was a large and overhanging one, and caused the owners much trouble in consequence of many of the stones being blown down, and others being decayed by the action of the acids emitted from the chimney. The whole of the cap was at last removed and built up with purposely-made fire-bricks. The firm estimate that the original cap in its erection and removal cost at least £700, and from their experience are convinced that no stone should be used at top, any overlapping to be gradually formed by hard burnt radiated fire-bricks.

MESSRS. ADAMS' SOAP WORKS CHIMNEY, SMETHWICK, NEAR BIRMINGHAM.

Designed by Mr. BUCKLE, Soho, Birmingham.

Description.—Circular brick shaft, erected 1835-6.

Dimensions,—

Height from foundation to top	. . . .	326′	10″
,, ground ,,	. . . .	312′	0″
Outside diam. at ground surface .	. . . .	27′	2″
Inside ,, ,, ,,	. . . .	15′	2″
Outside ,, top .	. . . .	5′	6″
Inside ,, ,,	. . . .	4′	0″

Weight.—Brickwork, 2,000 tons; concrete, sand and lime, 150 tons.

Bricks.—The firm state the bricks used were of large size, and weighed 4 tons per 1,000. The total number of bricks used in shaft was 500,000.

Building Operations.—The chimney was commenced in the autumn 1835, and allowed to stand during the winter and was finished in Sept. or Oct., 1836. At the time of its erection this shaft was the highest in the kingdom. The builder who began the work felt alarmed when the shaft had been erected about half its height, and the firm had to finish it themselves.

Lightning Conductor.—This shaft has been five times struck by lightning, once during the building when the chimney had reached about 200' in height, and four times since its completion. No very serious damage was done but once, when perhaps the electric fluid was aided by the lime having been abstracted from the mortar by the action of the hydrochloric acid in the escaping gases. The owners were then compelled to take down a portion of the top. These four strokes all happened after the lightning rod had been eaten away at the top by the hydrochloric acid, the remainder of the rod, however, being perfect. The rod is formed of $\frac{1}{2}''$ iron, and being eaten down only a few feet from the top may have had the effect of diminishing the force of the lightning. The conductor has several times been renewed, but the action of the acid was so rapid that it has soon been destroyed at the top, where it is exposed to the action of the escaping gases. In consequence of the rod so soon being eaten away, the firm, about 1871, had the conductor coated with platinum, with very favourable results. A few years ago about 30' more of the shaft were removed from the top by Mr. Frith, builder, of Coventry, who commenced fixing his tackle for the purpose by kite-flying, the chimney being in use all the time.

The total height is now about 250' above ground.

Cost, £1,700.

MESSRS. P. DIXON & SONS COTTON FACTORY, SHADDONGATE, CARLISLE.

Figs. 11, 12, 13, 14 *and* 15.

Architect, R. TATTERSALL, Manchester; *Builder,* RICHARD WRIGHT, Carlisle.

Description.—Octagonal brick; built Sept. 11th, 1835—Oct. 25th, 1836.

Dimensions,—

Height from foundation to top	320'	6"
,, ground line	300'	0"
Outside measurement at ground line	17'	4"
Inside ,, ,, ...	9'	6"
Outside ,, at top	9'	0"
Inside ,, ,, ...	6'	8"

Foundation.—Concrete, 6' deep, 35' 6" diam.

The footings are 6' in height, built up in 4 courses of 1' 6" each, and having a set off each time of 1' 6". The bricks in the first course of footings numbered 1,997.

Construction.—From top of footings to ground line, 8' 6", the base of shaft is circular, being 19' in diam., with walls 5' thick. The flues enter the shaft in this circular base on four sides, and are each 6' 9" high × 2' 6" wide, and are lined with 9" fire brick. The shaft from the ground line is octagonal, and is built up in eight sections as follows:—

1st section	. . .	30'	0" high	3'	11" wide	= 5 bricks.
2nd ,,	. . .	60'	0" ,,	3'	6" ,,	= 4½ ,,
3rd ,,	. . .	30'	0" ,,	3'	2" ,,	= 4 ,,
4th ,,	. . .	30'	0" ,,	2'	9" ,,	= 3½ ,,
5th ,,	. . .	30'	0" ,,	2'	4" ,,	= 3 ,,
6th ,,	. . .	30'	0" ,,	1'	11" ,,	= 2½ ,,
7th ,,	. . .	30'	0" ,,	1'	7" ,,	= 2 ,,
8th ,,	. . .	60'	0" ,,	1'	2" ,,	= 1½ ,,

300' 0" from ground line.

Batter.—1 in 72.

Cap.—The cap is of stone 7' 0" in depth, projecting 3' 0"; above this there is a blocking course of brick 8' 3" in height of 14" work.

Scaffold.—The chimney was erected by internal scaffold, stages being erected as the work proceeded. The men and materials were hoisted in boxes, purposely constructed, by a crab worked by 4 men.

Draught.—The draught, as ascertained by Mr. Hugh U. McKie, City Surveyor, Carlisle, in Sept., 1877, was equal to a column of water $1\frac{3}{10}$" in height. This observation was taken among others to find the influence of the draught upon sewers which Messrs. Dixons had allowed the Carlisle Town Council to connect with their chimney, and it was found the sewers were "perceptibly ventilated for a radius of four hundred yards, equal to an area of 502,656 square yards, or over 103 acres."

Duty.—Four boilers connected to shaft.

MESSRS. MITCHELL BROTHERS, MANCHESTER ROAD, BRADFORD.

Architect, Mr. MARK BRAYSHAW; *Builders*, Messrs. JOHN MOULSON & SONS.

Description.—Octagonal stone.

Dimensions,—

Height from foundation to top . . . . 330'
    ,,   ground   ,, ... 300'
Flue, perpendicular throughout . . . . 7'
External diameter at foundation . . . . 20'
  ,,   ,, at top   . . . . 9'

Concrete Foundation,—

1 course 22 × 22 × 1'
1 ,,   21 × 21 × 1'

MESSRS. J. CROSSLEY & SONS, DEAN CLOUGH MILLS, HALIFAX.

Architect, Messrs. R. IVES & SONS, Halifax; *Builders,* Messrs. PICKLES BROS. Built in 1857-8. Time occupied, 18 months.

Description.—Octagonal stone shaft, with circular flue.

Dimensions,—

Height from foundation to top . . . . 330'
  ,,   ground line . . . . 300'
Outside measurement at foundation . . . . 32'
Inside   ,,   ,, ... 9'
Outside   ,,   at ground surface. . . . . 30'
Inside   ,,   ,, ... 9'
Outside   ,,   at top . . . . 15½'
Inside   ,,   ,, ... 9'

Duty.—To carry off the smoke from 15 boilers.

Foundation.—The shaft is founded on rock.

Scaffold.—Inside.

Fire-brick Lining.—There is an inner circular shaft of 14" fire-brick, with a space of 3" between it and main shaft.

Weight, 8,300 tons.

Cost, £9-10,000.

MESSRS. I. C. JOHNSON & CO.'S CHIMNEY CEMENT WORKS, GREENHITHE.

Fig. 16.

Architect, I. C. JOHNSON; *Builder,* JOS. BLACKBURN.

Description.—Circular brick.

Dimensions,—

Total height, including foundation	. . . .	304′	0″
Height from ground line	. . . .	297′	0″
Outside measurement, foundation	. . . .	30′	0″
Outside ,, at ground line sq.	. . . .	25′	0″
Inside ,, ,, ,, cir.	. . . .	17′	6″
Outside ,, at top ,,	. . . .	11′	0″
Inside ,, ,, ,,	. . . .	8′	9″

Duty.—The smoke and vapour from 19 cement kilns.

Fire-brick.—No fire-brick lining, the chimney being 300′ from kilns.

Foundation.—The concrete foundation is 30′ sq. × 4′ thick, laid on the solid block chalk. The brick footings are 3′ in height.

Bricks.—Burham gault, No. 3 wire cut, 9″ × 4½″ × 2¾″ full.

Shaft.—The bond is Flemish. The shaft for the height of 10′ from footings is 25′ square externally, with a circular internal diameter of 17′ 6″. There are constructed at the base for a height of 10′ four walls or withes to prevent counter draughts.

There are four openings made at base of shaft, one on each side of the square, for although only two openings were required, Mr. Johnson advisedly constructed four, so that the brickwork should be equal in strength on all sides; the openings not wanted are built up in 9″ work.

In one of the openings there is a small furnace, which is only required at starting fire.

The shaft is constructed in eight sections as follows :—

1st section	. . . .	10′ high sq.	5 bricks thick.
2nd ,,	. . . .	41′ ,, circ.	4½ ,,
3rd ,,	. . . .	41′ ,, ,,	4 ,,
4th ,,	. . . .	41′ ,, ,,	3½ ,,
5th ,,	. . . .	41′ ,, ,,	3 ,,
6th ,,	. . . .	41′ ,, ,,	2½ ,,
7th ,,	. . . .	41′ ,, ,,	2 ,,
8th ,,	. . . .	41′ ,, ,,	1½ ,,

297′ high above footings.

The upper 41′ being of smaller diameter than the others, the bricks did not quite fit into the circle, so the tips were cut off the stretchers; a 9″ brick suited the circumference everywhere else.

The shaft stands on ground about 50′ above river level at high water.

Scaffold and Construction.—This shaft was built up from inside. Battens on edge were built in every 5' or 6', and covered with boards to form a platform for men to work upon. When the top was reached these battens were cut out and dropped down, and as they were available for other purposes the cost of scaffold was comparatively trifling. The materials and men were hoisted, by a crab and good manilla rope, up the centre of the scaffold, and not a single accident happened during the whole time of building.

Cap.—A string course 20' from top is made of three salient courses of brick, and 5' from top four courses oversail to about 9". The top salient course of same is of stone 3" thick and was put together in segments clamped with galvanized iron clamps run with lead.

Lightning Conductor. — $\frac{5}{8}$" copper rope, fixed by Messrs. Newall & Co., as follows:—One of the bricklayers was lowered outside by a rope, before the internal scaffold was struck; he fixed gun metal eye bolts into the brickwork at intervals, passing the wire rope (which had been drawn up inside) downwards through them until the whole was fixed.

MERRIMACK MANUFACTURING COMPANY, LOWELL, MASS., U.S.A.

Engineer, J. T. BAKER, C.E.; *Built,* 1882.

Description.—Circular brick shaft, with inner shaft and core.

Dimensions,—

Height above ground line	282'	0"
Outside diam. foundation	30'	0"
Outside ,, 2' 6" above ground	28'	0"
Inside ,, ,, ,, ...	12'	0"
Inside ,, at top	12'	0"

Foundation.—The chimney is founded on a ledge of sandstone. The foundation, 30' in diam., is built of granite blocks, laid on their natural beds. At the surface of the ground there is a dressed granite base 2' 6" in height, laid in clear Portland cement, the remainder of the foundation being in Rosendale cement and sand.

Upon this base is placed the brickwork, consisting of three cylinders, as follows:—

Outer Shaft.—Batter, ·42" per ft. for a height of 100'.

1st section	75½' high, 28' diam. 24" thick.
„	At junction of inner shaft, 36½" „
2nd „	60' high. . . .	20" „
3rd „	70' „	16" „
4th „	74' „ including cap 12" „

279½' high above granite base.

Inner Shaft.—Vertical, 18' diameter; 75½' high; 8" thick.

At this height the inner shaft connects with the exterior brickwork, making the masonry at that point 36½" thick, as above.

Lining or Core.—Uniform inside diameter, 12'.

It is entirely separate from the outside masonry, except the doorways and flue openings, and is built up as follows:—

1st section	100' high	16" thick.
2nd „	60' „	12" „
3rd „	90' „	8" „
4th „	29½' „	4" „

279½' high above granite base.

Construction.—The core was laid in mortar of lime and sand; the outside shell in lime, cement and sand.

Ladder and Lightning Conductor. — On one side of the chimney is a ladder of iron extending from the ground to the top, and on the opposite side is a $\frac{3}{4}$" galvanised iron wire rope, both ladder and rope being connected with a copper ring, having four spurs, the central point of which extends 8' above the top of the shaft. The bottoms of both ladder and rope are connected to a 16" water pipe.

Duty.—Two wrought-iron flues enter the chimney, one 5' × 6', and the other 5' × 11'. The chimney is constructed to provide for 15 sets of boilers; only 12 are now in use. Each set has 103¼ square ft. of grate surface, and is rated at 300 h.p.

Weight.—Chimney, 3,392 tons; cap, 18,600-lbs.

Materials.—1,101,000 bricks; 6,875 cubic feet stone masonry.

Cost.—18,500 dollars, or £3,854. 3s. 4d.

MESSRS. COX BROTHERS, CAMPERDOWN LINEN WORKS,
LOCHEE, DUNDEE.
Fig. 17.

Architect . . . . Mr. JAS. MACLAREN.

Description.—Ornamental brick, square to a height of 230′.
A balcony or cornice is here constructed. From this to top the
chimney is octagonal.

Dimensions,—

Height from foundation to top	. . . .	296′	0″			
,, ground line	. . . .	282′	0″			
Foundations		35′	0″ square, walls	12′	0″ thick.
At ground line		30′	0″ ,,	,,	6′	0″ ,,
At top of 1st panel		24′	0″ ,,	,,	3′	3″ ,,
,, main ,, ...		21′	3″ ,,	,,	2′	5″ ,,
,, balcony		20′	3″ ,,	,,	2′	0½″ ,,
,, chimney		19′	0″ ,,	,,	1′	6″ ,,
Circular flue at base		14′	6″ internal diam.	1′	6″ ,,
,, ,, top		13′	8″ ,,	,,	0′	9″ ,,

Construction.—The chimney is panelled and ornamented
by designs in parti-coloured bricks. The base is of ashlar, and
surmounted by massive stone mouldings. Above this is the
first panel, the pilasters of which are checkered red and white.
Above this is a base moulding, out of which spring the sides of
the next or main panel, which extends to a height of 185′ 2″
above the ground. It is relieved in the centre by loop holes
and sham clock holes at the top, and the sides of the main
panels are striped red and white. The tops of the main panels
are arched, and above them are two smaller panels on each face
of the chimney, surmounted by a Grecian frieze and other
ornaments in white and black bricks. Over this is constructed
the cornice or balcony, round which there is an ornamental
iron railing.

Ladder to Balcony.—The outside walls being square, while
the inner shaft is circular, a space is left at each corner, and
access to the balcony has been obtained by utilizing one of
these spaces in which to construct an iron ladder.

Flues.—The flues are of elliptical form, 9′ 6″ × 5′ 6″, and
constructed in each of the four sides, but at present only two
are in use.

Duty.—There are 58 furnaces connected to the chimney,
and also 13 smith's forges, and in addition the draught is used
for other purposes.

Draught.—The draught is now equal to a water pressure
of 1$\frac{8}{10}$′.

WROUGHT IRON CHIMNEYS.

Wrought iron shafts have found great favour in America and Russia, but in England and the Continent generally, as far as we have been able to ascertain, they are an exception.

In addition to the wrought iron shafts, detailed descriptions of which will be found in order of their respective heights (see Index), we have been informed of the following:—

Messrs. Witherow & Gordon, of Pittsburg, Pennsylvania, U.S.A., have, since 1876, built upwards of 30 wrought iron shafts, varying in height from 100' to 190', and from 5' to 9' in diameter. The firm write us that these shafts answer admirably the purpose for which they were built.

Mr. L. S. Bent, Superintendent of the Pennsylvania Steel Company, Steelton, Pa., U.S.A., states that his Company have the following eight wrought iron shafts in use, and have found them both durable and economical:—

No. 1	. . .	170' high	6'	6" diam.,	built	1881
1	. . .	165' ,,	6'	6" ,,	,,	1877
1	. . .	135' ,,	7'	0" ,,	,,	1880
1	. . .	112' ,,	6'	0" ,,	,,	1881
4	. . .	110' ,,	7'	0" ,,	,,	1869-74-5-6

They are lined for 30' with 9" fire-brick, and the remainder of height with 4" red bricks.

M. M. Schneider & Co., Creusot.

At the above works, in 1869, the chimneys that had served for nine years to carry away the products of combustion from 24 boilers became inadequate to meet the increased requirements of the establishment, which had been considerably extended from time to time; and it was decided to construct a new chimney in connection with an additional group of boilers. This new shaft it was determined to build in iron. A W.I. shaft had already been erected at the works 197' in height, 4' 3" in diam. at top, and 10' diam. at bottom, with plates from $\frac{3}{32}$" at top to $\frac{7}{10}$" at base in thickness, and weighing 28 tons, and had been constructed on the ground, and successfully raised in bulk to its assigned position. This formed a guide to the planning and construction of the proposed new and much larger shaft, the details of which are as follows:—

Constructors, M. M. Schneider & Cie; *Engineer,* M. Geay.

Dimensions,—

Height of masonry foundation above ground 1 M = ... 3' 3¾"
Height of iron shaft from top of masonry foundation 84·35 M = 276' 8¾"

Total height from ground line . . . . . . 280' 0"
Outside diam. of masonry 1 M, above ground line 7 M = . . . 22' 11½"
    ,,   ,,   shaft at top 2·30 M = ... 7' 6¼"

Construction.—The base of shaft is in form a frustrum of a cone to a height of 10 m., or 32' 8", and is built up of eight rings of plates each 1·25 m. = 4' 1" high, measuring from C to C of rivets. The base is fixed to the masonry foundation by a very strong ∠ iron ring riveted to chimney plates, and secured to foundation by holding-down bolts.

At the 9th ring, from bottom of shaft, the circumference is made up of eight plates 14 mill. = ·55104" in thickness, and at the upper part the rings are each composed of four plates only, 7 mill. = ·27552" in thickness.

Scaffold.—This shaft was erected by means of a "flying scaffold," that is cross-arms or bearers resting on angle irons, riveted to the inside of the wrought iron plates as erected. These bearers carried an internal platform, and an iron tube with cross-head timbers at top, from which was suspended an outside platform, so that the men could work both inside and outside the shaft. The scaffold was raised each time a complete ring of plates had been riveted up. This was done by two beams being placed across the top of the completed ring of plates, each beam being provided with two large nuts, through which screw rods worked. Near the bottoms of the four screw rods ratchet wheels were fixed, and the four ratchet wheels were worked simultaneously, until the whole scaffold had been raised the required height, and the cross arms or bearers brought up to the level of the next set of internal angle irons, to which the scaffold was secured. The next ring of plates was then commenced.

Ladder.—The internal ∠ irons used in erection remained in the shaft, and would serve for future ascension.

Fire-brick Lining. — An inner lining of fire-bricks was constructed for a height of eight plates, or 32' 8".

Weight. — Masonry in foundation, about 300 tons; weight of iron shaft, 80 tons.

Cost.—This shaft, exclusive of foundation, cost £1,600.

E

BARROW HÆMATITE IRON AND STEEL WORKS, BARROW-
IN-FURNESS.

Architect, A. WORRALL; *Builder*, A. J. WOODHOUSE.
Built, 1865—May to Sept. inc.—5 months.

Description.—Circular brick.

Dimensions,—

Total height, including foundation	. . . .	282'	0"
Height from ground line to top	. . . .	259'	0"
Outside measurement at foundation	. . . .	41'	8"
Inside	,, at base of shaft	16'	8"
Outside	,, at ground line	31'	0"
Inside	,, ,, ,,	16'	8"
Outside	,, at top	16'	6"
Inside	,, ,,	15'	0"

Foundation Bed.—Stiff clay.

Fire-brick Lining.—The shaft is lined throughout with 4½"
fire-brick, bonded to main shaft every third course.

Scaffold.—Inside.

Brickwork.—Purposely-made bricks were used in the construction laid to Old English bond, without hoop iron.

Lightning Conductor.—Copper stranded rope.

AMOSKEAG MANUFACTURING COMPANY, MANCHESTER, NEW
HAMPSHIRE, U.S.A.

Architect, GEO. W. STEVENS; *Steam Engineer*, CHAS. H. MANNING.
Builders, AMOSKEAG MANUFACTURING COMPANY.

Description.—Circular brick; built 1883; 60 days occupied
in construction.

Dimensions,—

Total height, including foundations	. . . .	265'	0"
Height from ground line to top	. . . .	255'	0"
Outside measurement at foundation	. . . .	25'	8"
Inside	,, ,,	19'	8"
Outside diam. at ground surface	. . . .	25'	0"
Inside	,, ,,	19'	8"
Thickness of brickwork ,,	. . . .	2'	8"
Outside diam. at top (exclusive of cornice)	. . . . .	12'	6"
Inside	,, ,,	10'	0"
Thickness of brickwork at top	. . . .	1'	3"

Foundation.—The shaft is founded on a bed of ledge. No concrete used.

Pressure.—On foundation (as given by the firm) is 10,220-lbs. per square foot.

Bond.—Headers every tenth course.

Batter.—1 in 40·8.

Bricks.—1,000,000 common bricks used in construction.

Weight.—2,330 tons.

Scaffold.—Outside scaffold used, costing 750 dollars = £156. 5s.

Duty.—This shaft was designed to burn 18,000-lbs. of anthracite coal per hour. It carries off the fumes from sixty-four boilers = 8,400 h.p. The company chiefly manufacture ginghams, tickings and fancy shirtings.

Inner Shaft.—The chimney has an inner shaft of 10' internal diameter.

Lightning Conductor.—Wrought iron; costing 95 dollars = £19. 15s. 10d.

Cost.—Complete, 16,000 dollars = £3,333. 6s. 8d.

MESSRS. LISTER & CO., MANNINGHAM MILLS, BRADFORD.

Fig. 18 *(the tallest chimney in Bradford).*

Architects, Messrs. ANDREWS & PEPPER; *Clerk of Works,* A. RHODES. *Builders,* Messrs. J. & W. BEANLAND.

Description.—Square ornamental stone chimney, of uniform width from base to top, with panelled sides.

Dimensions,—

Height from ground line to top	. . . .	256'	6"
Outside measurement base and top	. . . .	21'	0"
Inside	„	at base	10'	0"
„	„	under cap	11'	0"
„	„	at top	13'	0"

Fire-brick Lining.—An inner shaft of fire-brick is constructed for a height of 50' in 9" work, leaving a cavity between it and the shaft proper of 4".

Foundation.—The foundation bed is clay; upon this is a layer of Lias lime concrete, 4′ thick × 40′ square, then two courses of large rag footings or landings, each course 12″ thick, well bedded and bonded. Some of the landings were 12′ × 7′ × 1′ in size. The quantity of stone used in the footings was about 3,300 cubic feet.

Lime.—Barrow Lias lime used throughout.

Materials.—Total weight, about 8,000 tons.

Scaffold.—Inside.

Cost.—£10,000 about.

WEST CUMBERLAND HÆMATITE IRON WORKS CHIMNEY.

Figs. 19 *and* 20.

Engineer . . . . . Prof. W. J. M. RANKINE.

Description.—Circular brick, Old English bond.

Dimensions,—

Total height, including foundation . . . . 267′ 0″
Height from ground line to top . . . . 250′ 0″
Foundation—bed of concrete 3′ deep × . . . . 34′ 6″ sq.
Outside measurement at bottom of footings . . . . . 31′ 6″
    ,,   ,, of sq. basement . . . . . . 30′ 0″
    ,,   ,, of bottom of circular shaft . . . 25′ 7″
Inside diam. ,, of circular shaft 2′ above oct. base. 21′ 10″
  ,,   ,,   ,, top of shaft . . . . 13′ 0″
Outside ,,   ,,   ,,   ,, ... 15′ 3″

Contract and Execution. — Tenders were invited from a limited number of builders in the north of England and in Scotland, and the lowest was accepted, being that from Messrs. William Wilson & Son, of Glasgow.

The progress of the building was restricted by the specification to a rate not exceeding 6′ vertical height per day.

Concrete Foundation.—In order that the concrete foundation might have time to harden, before being subjected to a heavy load, it was made by the Iron Company themselves before the contract for the chimney was let.

Building.—By dimensions given above and reference to diagrams 19 and 20, it will be seen the concrete foundation is square. Upon this bed four courses of footing are built, then a basement 30' square, which is gathered into an octagon by gradually stepping the brickwork at the corners; from this line to the top the shaft is circular.

There are four circular openings, 7' 6" diameter, with archings three bricks thick in the base for flues.

The batter of circular shaft is uniform throughout, and was adopted because the accuracy of building can be tested at any moment by the eye without the aid of instruments.

Thickness of Brickwork,—

> 1st section at top 80' 0" = 1½ bricks, including fire-brick lining.
> 2nd ,,   80' 0" = 2   ,,   ,,   ,,
> 3rd ,,   88' 0" = 2½ ,,   ,,   ,,
>   2' 0" = stepped from 2½ to 4 bricks.
>
> Total, 250' 0" from ground line.

Scaffolding.—Internal scaffolding was used, and in its construction great care was taken that the horizontal beams should be wholly supported by the brickwork, and not by the vertical posts, for great danger has been known to arise from the upper brickwork coming to bear upon the ends of the horizontal timbers, and through them on to the vertical posts, owing to the settlement of the lower part of the chimney. (See description of Port Dundas Chimney—"Straightening.")

Stability.—Prof. Rankine gives the *bed joint of least stability* at 2' above the ground line, and the deviation of the resultant pressure from the axis of the chimney at that joint, which would be produced by such a wind as 55-lbs. per square foot is 6' 4", being a fraction of an inch less than ¼ of the outside diameter.

Pressures.—The following are the intensities of the mean pressures due to the load on different bed joints:—

> At 2' above the ground line . . . . 8 tons on the sq. ft.
> In basement at springing of the arches . . 3   ,,   ,,   ,,
> On upper surface of concrete 2   ,,   ,,   ,,
> On ground below   ,, ... 1·6 ,,   ,,   ,,

Fire-brick Lining.—This is included in the thickness of brickwork, as before stated :—

> Upper 160' 0" of the shaft . . . . ½ brick thick.
> Lower part of the cone, basement, flues, and archways . . 1   ,,

The fire-brick lining is bonded with the common brickwork in the ordinary way, the only difference being that the fire bricks are laid in fire-clay and the ordinary bricks in mortar. The reasons given for adopting this mode of construction, by

Prof. Rankine, in preference to an internal fire-brick chimney, are as follows:—

1st. When the fire-bricks are bonded with the ordinary bricks they contribute together to the stability of the chimney, and so save an additional thickness of ordinary brickwork.

2nd. Unless the internal chimney is carried up to the top of outer cone there is a risk of damage through the explosion of gaseous mixtures in the space between.

3rd. There is also a risk of the cracking of the outer cone at and near the upper end of the inner cone, through unequal heating at that place, unless the inner shaft is carried to the top of the outer one.

The basement is paved inside with 6″ of fire-brick, resting on 6″ of common brick laid on the concrete.

Ordinary Brickwork.—The ordinary brickwork is built of white bricks of good quality supplied by the Iron Company. The bond is Old English. In the basement there is one course of headers to every two courses of stretchers. In the cone one course of headers to every three courses of stretchers.

Hoop Iron.—Strips of No. 15 B.W.G. hoop iron, tarred and sanded, are laid in the bed joints of the cone, at intervals of 4′ in height, with their ends turned down into the side joints. Care was taken to bed the hoop iron on the common brickwork, and not on the fire-brick lining. The length of hoop iron, in each bed joint in which it is laid, is twice the circumference of the chimney at that point.

Mortar.—In the concrete foundation, the basement and a small part of the cone, the mortar was made of hydraulic lime. Owing to an unexpected difficulty in obtaining such lime on the spot, it had to be brought from a distance, at considerable expense, and, therefore, the mortar for the rest of the building was made of a very pure lime from the immediate neighbourhood, rendered artificially hydraulic by a mixture of iron scale from the rolling mills at the works. The following are the proportions by measure:—

Lime	2
Scale	1
Sand	5
Total	8

The use of iron scale for hardening mortar and making it artificially hydraulic should be more generally known.

Cast Iron Curb.—On the top of the chimney is a cast iron curb, 1″ thick, coming down 3″ both inside and outside. It was "paid" over with a coating of pitch when fixed.

Lightning Conductor.—The lightning conductor is of copper wire rope, about $\frac{3}{4}$″ diam. It terminates in a covered drain.

Cost.—Actual cost, including designing and superintendence, was £1,560, being at the rate of almost 4d. (fourpence) per cubic foot of the whole space occupied by the building, which is 94,000 cubic feet nearly.

Duty.—This chimney has to carry off the gaseous products of combustion from four blast furnaces, and from various stoves and boilers that are heated partly by burning the inflammable gas from the blast furnaces and partly by coal.

The total quantity of solid fuel consumed is about 10¼ tons per hour, when all the furnaces are at work.

Temperature and Draught. — The temperature inside the chimney, when doing about three-quarters its full duty is 490° Fah., and the pressure of the draught is 1⅛″ of water, which agrees to a very small fraction with the pressure as deducted theoretically from the temperature and the height of the chimney.

LANCASTER: MESSRS. STOREY BROTHERS & CO., WHITE CROSS STREET MILLS.

Fig. 21.

Engineer, Mr. EDWARD STOREY; *Architects,* Messrs. PALEY & AUSTIN.

Contractor, Mr. C. BAYNES.

Built 1876-7-8; about 18 months being occupied in its construction.

Description of Shaft.—Octagon brick with stone cap.

Dimensions,—

Total height, including foundation	270′	0″
Height from ground line to top	250′	0″
Outside measurement at foundation	28′	0″
Inside ,, ,, ...	17′	0″
Outside ,, at ground surface	25′	0″
Inside ,, ,, ...	17′	0″
Outside ,, at top	10′	8″
Inside ,, ,, ...	9′	2″

Materials.—750,000 brick, 650 cubic feet stone and 145 cubic yards concrete.

Foundation.—The shaft stands upon a base of concrete 28′ square × 5′ thick.

Bricks.—"Shale" bricks, supplied by the Caton Brick and Tube Company, were used. These were chosen because they are said to absorb less moisture than ordinary bricks.

Internal Shaft.—This is octagonal in form, 264′ high × 8′ internal diameter, built 18″ thick at base and 9″ at top. It is built parallel to within 12′ of the top, then sets back 7″ at each side, as shown in diagram 21. The inner shaft carries off the smoke from the steam boilers. It is surrounded by a space or cavity 2′ 6″ wide, enclosed by the outer shaft, and the vapours from the stoves, &c., are passed off through this 2′ 6″ space. This space between the inner and outer shells is divided into three distinct flues by vertical diaphragms of brickwork, which latter serve to tie the whole structure together.

Outer Shaft.—The brickwork of outer shaft is 4′ 6′ thick, where it rests on the base, and 14″ thick at top. About 20′ from top of chimney the outside wall curves inwards, and joins the inside wall as shown on drawing.

Annular Flue or Cavity.—It will be seen at the junction of the inner and outer shafts two outlets are constructed in each of the eight sides, so as to allow the heated air, vapours, &c., to escape from the annular flue, the principal of which is more fully described under the head of Print Works Chimney, Falls River, U.S.A.

Weight.—Total, 3,300 tons.

Cost.—£2,800 complete.

Deflection and Straightening.—During the erection of this chimney, when it had attained a considerable elevation, it canted out of the perpendicular towards the south 3′ 10½″ at top. This was accounted for partly because during the two years occupied in its erection very frequent rains kept the mortar soft on the weather side, the result being that the joints on that side were squeezed rather closer than those on the other, and the stalk heaved or bent over. The principal reason, however, must have been that the foundation was a little weak on the south side, and thus yielded to the pressure, the weight on that side being increased by the deflection of the chimney.

The work of bringing it back plumb was successfully executed by Mr. J. W. Cronshaw, of Blackburn. The operation consisted in cutting out courses of bricks in 5 different places near the base on the north side of the chimney, and rather more than half way across. These courses were replaced by ones diminishing very slightly in thickness from south to north, so that the five courses shortened the north side of the chimney sufficiently to bring the axis in a true vertical position. The process of cutting was as follows :—

A width of about 18″ was first cut right through the 4′ 6″ brickwork of the outer shell on the extreme north side, a course of bricks being then withdrawn, and the top and bottom joints being thoroughly cleaned off. Into this space 18″ wide, good hard bricks were then closely packed, these bricks being thinner, as already explained, than those they replaced. At the outer end good oak wedges were inserted; then similar cuts were made right and left of the first, and similarly treated with bricks and wedges, the work proceeding regularly on each side, east and west, from the north towards the south. As this was done the wedges were gradually withdrawn, and the chimney quietly settled over towards the north, until when all the five cuts were completed it had come back to the perpendicular.

The lowest cut was close to the ground, and the highest about 30′ above. Two of the cuts were continued into the internal shaft.

CONNAH'S QUAY CHEMICAL COMPANY'S CHIMNEY.

Description.—Square brick shaft.

Dimensions,—

Total height from foundation to top	258′	6″
Height from ground line to top	245′	0″
Outside measurement at foundation	28′	3″
Inside „ at ground surface	17′	6″
„ „ at top	7′	0″

Weight and Materials.—The weight of stone in this erection was 645 tons, in addition to which 1,078,000 bricks were used.

Cost.—£2,000. This price will doubtless be considered very low, but the Connah's Quay Chemical Company, for whom this stalk was erected, say that the cost was but little over the above sum.

NEWLAND'S MILL CHIMNEY, BRADFORD.

Figs. 22 to 30 inclusive.

Built for the late Sir H. W. Ripley, Bart., 1862-3, by Messrs. JOHN MOULSON & SONS.

Description.—Octagonal stone shaft.

Dimensions,—

Total height from top of concrete shafts	260'	0"
Height from ground line	240'	0"
Concrete foundation, sq.	32'	6"
Outside measurement on stone footings	24'	0"
Inside diam.	12'	2"
Outside measurement at ground line	24'	0"
Inside diam. ,,	12'	2"
Outside measurement under cap at top	14'	0"
Inside diam. ,, ,,	9'	0"

Fire-brick Lining.—9" thick.

Height from top of footings	30'	0"
Inside diam. throughout	9'	0"
Withe, height from footings	10'	0"

Fig. 25.

Foundation.—The site was an old coal shaft, which was filled up with Skipton lime concrete, forming a centre pillar 8' 6" diameter. Round this were constructed four other shafts, each 6' diameter, also of concrete, the sinking of which cost 9s. 6d. per yard per shaft. Upon this formation was laid a tabling of lime concrete 32' 6" square × 2' 6" thick, the base courses or footings of the chimney resting upon it. The concrete was not rammed, but was tipped in from staging; it was almost liquid and nearly levelled itself by the drop.

The old workings surrounding the centre shaft were packed with stones and oak wedges.

The amount of contract for sinking the four shafts, packing coal beds, cleansing and searching old workings, and ascertaining condition of ground, was £104. 17s. 9d.

Figs. 24, 29, 30.

Materials. — Outer stone casing, stone backing, and 9" common brick lining throughout, fire-brick at bottom.

Weight.—Total, 3,600 tons; above cuts (made in straightening described further on), 2,230 tons.

Pressure,—

4·5 tons per foot sup. of foundation.
22·4 ,,   ,,   on 5 concrete piers.

Contract.—The construction of the chimney was settled as follows:—The late Sir H. W. Ripley, Bart., being desirous of erecting a chimney, sent for Messrs. J. Moulson & Sons in May, 1862, to give a tender. There were no plans or specifications prepared when the tender was given, but the following formed the basis of the estimate:—Chimney to be 80-yds. high, 9' flue, base 24' square, with two courses of footings 12" thick, the first 28' square, the second 24' square, placed on a good bed of concrete. The builders undertook to execute the work for £942. 5s. 10d. The Ashlar foundations, concrete and capping, to be extra.

Building.—At the ground line the chimney was a regular octagon, and was so continued upwards at a regular batter of $\frac{7}{8}$" to the yard.

The chimney was commenced July, 1862, and continued to the middle of December (the back end of the year being open), the building was then suspended, being a little more than 120' high. The work re-commenced 28th February, 1863. "Through" stones were built in at about every 3' in circumference, and about 2' 3" apart vertically. The erection was continued to the 7th June, the chimney having reached 210' in height.

Figs. 29 and 30.

Straightening.—On the evening of this day (7th June), the chimney was left plumb. On the 8th it was found to be bulged on one side, and hollow on the other. About 54' from ground line a course of stones was cut out on the opposite side to the canting over. Two men outside with long chisels cut away, say for 1' wide on the outside, a 7" stone course and through the backing; two men from the inside cut through the brick lining, and met the opening cut through the backing from the outside. This space was filled up with stones $\frac{1}{4}$" less in thickness; upon these were placed long feather-edged iron wedges to make up the original thickness. This operation was continued nearly half-way round, with the exception of the angles which were left. Before proceeding to draw the wedges, mastic cement was introduced by means of syringes. The wedges were then withdrawn by hammering them sideways, both from inside and outside. As the wedges were removed the men could hear the "through" stones breaking. The first cut did not have the desired effect, and a second was decided upon about 2' above the first, with the same results as to breaking the "throughs." The chimney was, after this, declared to be perpendicular. In the "coming to," the corner stones at the angles were crushed for about 12' above and below the cuttings. These were replaced, and the chimney was then completed.

Cracks.—Three years after completion the chimney was found to be cracked and broken on the side opposite to the cuts. This was repaired at a cost of £96.

About 1872 further cracks were noticed, which were repaired.

In October, 1882, the tenants of the mill became uneasy about further indications of cracking, which, in December, developed into bulges. Upon examination it was decided to take out the bulges and repair the outer casing, it being the general opinion the latter was alone at fault. Difficulty was experienced in this, and the attempt to rectify the bulge failed.

Collapse.—On the 26th Dec. small portions of the outer casing fell, and on the 27th a large piece fell, breaking down the scaffold used for the repairs. On the night of the 27th the wind blew half a gale, or about 16-lbs. per foot super. On the following morning, 28th Dec., more of the outer casing fell, and at a few minutes past eight a.m. the chimney began to settle, bursting out stones and lime near where the chimney had been cut. This continued for a few seconds, then the upper portion of the chimney fell in a S.E. direction, killing 54 persons, and destroying property estimated at £20,000.

In 1884 a test action was brought against Messrs. Ripley by the sufferers through this disaster, and resulted in the whole ·matter being referred to an arbitration, which awarded in the aggregate £2,500 as damages.

PACIFIC MILLS, LAWRENCE, MASS., U.S.A.

Architect and Builder, H. F. MILLS, C.E.; *Built,* 1873.

Description.—Brick, octagonal outer shaft, circular inner shaft, vertical inner lining. Shaft situate 210′ from boilers.

Dimensions,—

Total height	242′	0″
Height of outer shaft, including footings	233′	0″
Height of inner lining	234′	0″
Outside measurement outer shaft at base	20′	0″
„ „ „ „ at top, under projecting cornice.	11′	6″
Inside diam. vertical flue	8′	6″

Foundation. — Foundation bed, 19′ below ground, coarse gravel,—

Concrete 35′ square, enclosed by pine sheet piling . . 1′ thick.
Rubble masonry of granite in Rosendale cement . . . 7′ high.

Outer Shaft.—This is constructed in six sections, viz.:—

1st section	12′ high	28″ thick.
2nd „	18′ „	24″ „
3rd „	20′ „	20″ „
4th „	40′ „	16″ „
5th „	60′ „	12″ „
6th „	83′ „	8″ „

233′ high above granite masonry.

Inner Shaft,—

1st section.	27' high	24" thick.
2nd „	154' „	12" „
	181'	

Lining,—

1st section	20' high	20" thick.
2nd „	17' „	16" „
3td „	52' „	12" „
4th „	145' „	8" „
	234' high above granite masonry.	

Construction. — The foundations were laid in mortar of Rosendale cement and sand ; the outer shell in mortar of Rosendale cement, lime and sand ; and the flue walls in mortar of lime and sand.

Duty. — In the winter of 1873, the vertical flue having reached 90' in height above ground, boilers having 452 square feet of grate surface were connected with the chimney, with satisfactory results. The chimney was designed to serve boilers having 700 square feet of grate surface.

Weight.—The approximate weight of the chimney is 2,250 long tons.

Bricks.—There were 550,000 bricks used in the construction of this shaft.

Lightning Conductor.—The shaft was struck by lightning in June, 1880, after which date a lightning rod was put up. It consists of a seamless copper tube, $\frac{5}{10}$" thick, 1" inside diameter, at the top of which are seven points radiating from a ball 4" in diameter, the top of the central point being 8½' above the iron cap. The rod is attached to the chimney by brass castings, and is connected at the base to a 4" iron pipe extending 60' to a canal.

DOVERCOURT CEMENT WORKS, MESSRS. JOHN PATTRICK AND SONS.

Figs. 31, 32 *and* 33.

Designed and built by Mr. JOSEPH BLACKBURN.
Built, 1883, between January and October ; 7 months occupied.

Description.—Brick, square base, octagonal shaft.

Dimensions,—

Total height, including foundations . . . . 249′ 0″
Height from ground line . . . . 230′ 0″
Outside measurement at base of footings . . . . 22′ 2″
Inside   „   at top   „   . . . . 9′ 2″
Outside   „   at ground line . . . . 19′ 3″
Inside   „   „   . . . . 8′ 9″
Outside   „   at top . . . . 9′ 5″
Inside   „   „   . . . . 7′ 0″

Foundation.—The foundation bed is clay; upon this is laid a block of concrete, 26′ square × 8′ thick, upon which the footings commence.

Fig. 31.

Pedestal.—The square pedestal is 37′ high above concrete bed, thickness of brickwork 5′ 3″, octagonal flue, without fire-brick lining. The foundation and pedestal were allowed time to settle before building of shaft was commenced.

Shaft.—Octagonal brick, thickness at base 4′ 3″, thickness at top 14½″. At the junction of the octagonal shaft and square pedestal pyramidal corners are constructed 12′ 6″ high, so as to equalise the bearing of shaft on square base.

Cap.—Constructed of brickwork.

Materials.—Concrete, six sand and gravel to one cement. Mortar, five river sand to one white chalk lime; to each ¼-yd. of mortar one bag of Portland cement added (the whole well tempered before use). Bricks, kiln burnt, 9″ × 4½″ × 2½″, with white Burnham brick angles specially made ; total, 280 m.

Weight.—About 1,800 tons.

Bond.—English.

Scaffold.—Inside, costing £7. 10s.

Lightning Conductor. — Copper tape, 2″ × ⅛″, terminated about 20′ from base of chimney, laid in carbonaceous material, well watered and rammed. Cost £40 fixed.

Duty.—This chimney serves twelve cement kilns and twenty-four coke ovens.

Cost.—Total, £2,000.

WOOLWICH ARSENAL, SHELL FOUNDRY, CHIMNEY.

Description.—Octagonal brick shaft, square base.

Dimensions,—

Height from foundation above concrete to top . . . . . 239′ 9″
Height from ground line to top . . . . 223′ 9″
Outside measurement of square base . . . . 20′ 0″
Height of base above ground . . . . 27′ 0″
   „   octagonal shaft above base . . . . 196′ 9″
External diam. of   „   at   „ . . . . 16′ 9″
   „   „   „   at top „ . . . . 6′ 6″

Thickness,—

Base of octagonal shaft . . . . . 2′ 7½″
Top   „   „   . . . . .   9″

The brickwork is reduced 4½″ at every 31′ 6″, the topmost length being 26′ in height.

Foundation.—Concrete.

Construction.—The whole of the chimney is built in mortar, with the exception of the top 9′, which is bell-mouthed, and built in cement. The total time occupied in the erection of this stalk was nineteen weeks.

Cap.—This is of Portland stone, with blocking course, and weighs about 17 tons.

Scaffold.—Inside.

MESSRS. J. C. GOSTLING & CO., CEMENT WORKS, NORTHFLEET, NEAR GRAVESEND.

Figs. 34, 35, 36.

Architect, JAMES CUBITT; *Builder,* Mr. BLAGBURN.
Built June, 1873; 16 weeks (good weather) occupied in erection. *Fell,* Oct. 2nd, 1873.

Description.—Circular brick.

Dimensions,—

Height, including foundation . . . . 227′ 3″
   „   above ground . . . . 220′ 0″
Outside measurement base of square footings . . . . 30′ 0″
   „   diameter at ground line . . . . 22′ 0″
Inside   „   „   „ . . . . 14′ 6″
Outside „   at top . . . . 11′ 0″
Inside   „   „ . . . . 9′ 6″
   „   „   at 4′ 7½″ from top . . . . 8′ 9″

Foundation Bed.—Chalk.

Shaft.—This was constructed in nine sections, viz.:

Footings	7' 0"		
1st section.	26' 3"	. . 3' 9" thick.
2nd „	26' 3"	. . 3' 4¾" „
3rd „	26' 3"	. . 3' 0" „
4th „	26' 3"	. . 2' 7¾" „
5th „	26' 3"	. . 2' 3" „
6th „	26' 3"	. . 1' 10½" „
7th „	26' 3"	. . 1' 6" „
8th „	31' 10½"	. . 1' 1½" „
9th „	4' 7½"	. . 9" „

227' 3" total height.

Batter.—1 in 40.

Cap.—This was formed of equal over-sailing courses of brick, laid in cement-mortar (see construction), each course projecting about $\frac{3}{16}$ of an inch beyond that below. The extreme projection attained at top of cap was 15½" beyond the vertical line drawn from starting point, but as it took 9' 9" in height to do this, and as the batter was 3" in 10', the projection of the top of cap from the receding line of shaft was 18½". Upon the cap there were eight flat projections or piers, constructed each 1' 11" wide, carried up in over-sailing courses, ranging with those of the body of the cap. The projection of these piers from top to bottom was 4½" beyond the main part of cap, and the bottoms of them were supported on a series of courses over-sailing more gradually than the upper ones. Above the cap the shaft wall was continued for a height of 4' 6" in 9" work.

Bond.—The bond was that known as " half-brick bond," and contained at least twice as many stretchers as would occur in Old English bond.

Weight.—Shaft, 1,674 tons; cap, 19 tons 3 cwt.

Pressure,—

On base of 3' 9" work	6¾ tons per ft. super.
„ 1' 1½" „	2¼ „ „
„ 1' 1½" „ below cap .	¾ „ „

Construction.—Messrs. J. C. Gostling & Co. made arrangements with Mr. Blagburn to provide labour for the erection of the shaft as above detailed for £500, the firm providing all materials. The best Dorking grey stone lime was used with the best Thames sand, every few courses being grouted in with Portland cement. The whole of the upper part of chimney was built of the best picked paviors, all imperfect bricks being rejected. The lower part of the walls was partly composed of the hardest stocks that could be obtained and partly of paviors ; and within 50' of the ground there was a small proportion—5%—

of rather over-burnt and somewhat vitrified bricks, approaching in character to "rough stocks." There was no difference between the facing and backing, the same quality of bricks being used for the entire thickness of wall. At intervals of about 3' two successive courses of brickwork were built with the vertical joints dry, and then grouted with neat Portland cement. The cement, which was thus poured in as grout, set admirably. After falling 200' these bricks were still found joined in double courses, and they had as often broken through their own substance as through the cement. There was not the least sign of expansion caused by the cement, and a large mill floor at the works was laid at the same time with the same lot of cement, and no flaw or blister subsequently appeared. The cement used was the best of the Burnham Cement Company's make. The original intention was to mix the cement with an equal measure of sand for the construction of the cap, but by the advice of persons experienced in this class of building it was used with a small quantity of mortar. The idea was, that though the ultimate strength of the cement and sand might be as great, yet the cement and mortar would adhere to the bricks better at the beginning. Whatever may have been the cause it appeared after the accident that a considerable part of this cement-mortar had not set with anything like the firmness of the neat cement grouting.

Collapse. — On Thursday afternoon, Oct. 2nd, 1873, the shaft being virtually finished, Messrs. Gostling attended to witness the laying of the last brick. At about one o'clock, when the workmen were about to ascend, one man having actually reached the top, the upper part of the shaft was observed to bulge outwardly, and immediately afterwards about 60' of the top fell, both outside and inside the chimney, resulting in the death of six and injury of eight men.

The outline of the fractured shaft was highest on the south-west side, and sloped irregularly in the opposite direction, as see dotted line, Fig. 34. The top of the ruin overhung considerably towards the north-east, and there were vertical fissures extending for a short distance down. The smaller fissures, and very likely some of the larger ones, were produced as follows:— The cross timbers, on which one after another the internal scaffolds had rested, were left in till the completion; at the moment of the accident great masses of brickwork fell on these timbers, and thus violently jarred the walls at the points where they were inserted, and the result was that many of these points were subsequently traceable on the outside by bulges and radiating cracks.

The only witness who deposed at the inquest that he saw the actual collapse from the outside stated that it began by

F

bulging at a point on the north-east side of the shaft some 10′ below the cap.

The remains of the chimney were blown down by the Royal Engineers on the Saturday afternoon following the accident. A charge of 5-lb. of gun cotton was first fired in the centre of the *debris*, in the inside of the chimney, to ascertain the effect produced on the loose masses at the top by the concussion. The cracks were opened and a few bricks brought down by it. A charge of 8-lb. of gun cotton was then placed in the centre of the chimney, about 20′ from the ground, by securing the charge to the end of a pole which was put through an opening in the side of the chimney from a high bank close to which the chimney had been built. No immediate result followed, but a few moments after the charge had been fired two large masses of brickwork fell from the top, and the cracks in the chimney opened very much and extended downwards. After waiting half-an-hour to avoid risk another similar charge was fired in the same position, when the top of the ruin fell, and the base crumbled to about 30′ from the ground. The remaining brickwork could now have been safely pulled down in the ordinary way, but to assist, four charges of 1½-lbs. each were placed in the cracks in the chimney and fired. The result, however, was only to open the cracks further, without bringing down any quantity of brickwork. The charges were fired by Professor Abel's detonators and Siemen's dynamo-electric machine.

Re-built. — The chimney was re-built in 1874, 220′ high, according to the original design. The special precautions which the accident in the previous year had shown to be indispensable were taken to ensure that all the bricks were wetted before being laid, and that none of the mortar or cement was worked up again after being spoilt or " killed." Mr. J. Cubitt acted as architect, Mr. Blagburn as contractor, and Messrs. Gostling as before supplied their own materials.

NEW YORK STEAM HEATING COMPANY, GREENWICH STREET BOILER HOUSE, U.S.A.

Engineer, Dr. CHARLES E. EMERY, M. Am. Soc. C.E.

Dimensions,—

Height above foundation	221′ 0″
„ „ high water	220′ 0″
„ „ basement floor	217′ 0″
„ „ grates of lower tier of boilers	201′ 0″
„ „ „ upper „ „	141′ 0″
Inside measurement	27′ 10″ × 8′ 4″

Foundation.—The beach of the Hudson River was, at some time, at this locality, and the foundation of the chimney was placed in fine clear beach sand, with some pockets of coarser sand and a little stone. The foundation is 1' below high water.

Construction.—It was necessary to place within a limited area a very large boiler capacity, viz., 16,000 h.p. This was done by making four stories of boilers; the chimney was, therefore, necessarily located with reference to these boilers, and the plan of the chimney was determined by the shape of the lot. The thickness of the walls on the interior of the building runs from 5' to 20", and on the other sides from 3' to 20".

Fuel, &c.—About 1,000 tons of coal will be burnt daily. It is expected that elevator arrangements will be perfected to receive this amount of coal each night. More trouble is experienced with the ashes than with the coal. Clearing is done every six hours. A new bar is used that turns on hinges and gives good results. Mr. Emery says:—"We have not made many experiments with coal dust; we have to use a fuel which has some reserve power to provide for possible contingencies. We find coal is worth about what is charged for it."

MESSRS. MARK OLDROYD & SON'S CHIMNEY SHAFT, KNOWN AS "BIG BEN," DEWSBURY.

Architects . . . . . Messrs. JOHN KIRK & SONS.

Description.—Round brick shaft without pedestal, built 1869.

Dimensions,—

Total height, including foundation	229'	0"
Height from ground line to top	210'	0"
Outside measurement concrete bed	40'	0"
,, ,, brick foundation	34'	0"
,, ,, at ground surface	23'	8"
Inside ,, ,, ,, ...	10'	0"
Outside ,, at top	12'	6"
Inside ,, ,, ...	10'	0"

Foundation Bed.—Gravel.

Batter.—1 in 37.

Bricks.—600 m. used in construction.

Weight.—2,000 tons.

Scaffold.—Inside.

Cap.—Built of stone.

Lightning Conductor.—Stranded copper rope.

Duty.—No. 8, 40 h.p. boilers connected to shaft.

Cost.—£1,200.

COLTNESS IRON WORKS, LANARKSHIRE.

Dimensions,—

Height above ground . . . . 210′ 0″
Outside measurement at base . . . . 18′ 6″
     ,,   ,,   at top . . . . 10′ 6″

Construction.—This chimney is built up in five sections, as follows:—

1st section . . . . . 35′ high 4½ bricks thick.
2nd ,, ... 40′ ,, 4 ,,
3rd ,, ... 50′ ,, 3 ,,
4th ,, ... 40′ ,, 2½ ,,
5th ,, ... 45′ ,, 1½ ,,

210′ high above ground.

Fire-brick Lining.—25′ high, 10″ thick.

CLEVELAND ROLLING MILL CO., CLEVELAND, OHIO, U.S.A.
Figs. 37 *and* 38.

Engineers and Constructors, Messrs. WITHEROW & GORDON, Pittsburg,
Pa., U.S.A.
Built Sept., 1881. About 50 days were occupied in its erection, apart from the building
of the foundation proper.

Description.—Wrought iron chimney, bell-shaped base.

Dimensions,—

Height, including foundations . . . . 213′ 6″
  ,,   from ground line to top . . . . 190′ 0″
  ,,   of bell-shaped base . . . . 21′ 0″
Outside measurement at foundation . . . . 30′ 6″
  ,,   diameter at foot of bell base . . . . 21′ 2″
  ,,   ,,   at top   ,, ... 13′ 6″
  ,,   ,,   ,, ... 12′ 0″
Internal diameter throughout . . . . 11′ 0″

Foundation.—Stone laid in cement, and is situate in what is termed the "Bottom" next to Cuyahoga River, where the ground is all of alluvial formation. For such a load as this chimney the foundation required close piling; piles were driven 23′ to 24′ in depth, and almost in contact with each other. Through the stone foundation, No. 8, 2⅛″ bolts were passed, connecting a circular cast-iron foundation plate of T section, 18″ × 8¼″ at bottom of stonework to a similar casting upon the top of stone foundation. This top circular ring or base plate is formed with a projecting flange placed at an angle of 60° to receive plates forming bell-shaped base, 2′ above ground.

Construction. — The chimney was constructed by inside scaffolding and built up one plate high at a time. The workmen hanging what is called a "cage" on the plates to serve as a stand for the "holder on" while riveting the plates *in situ.*

Bell-shaped Base.—The plates forming the base are bolted to the flange of chimney base ring by ¾″ bolts, and when completed to a height of 21′ form a bell-shaped base 21′ 2″ diameter at bottom and 13′ 6″ at top.

Shaft.—From the top of bell-shaped base the wrought-iron outer casing is continued to a height of 21′ from below top; from this point the cap is formed as shown on drawing.

Rivets and Riveting.—The plates are all riveted together with a lap of 2″.

The constructors used conical-shaped rivet heads, and the diameter of rivets for this class of work is as near as possible twice the thickness or upwards of plate, and the pitch of rivets is 5 diameters.

Ladder.—A wrought-iron ladder is fixed to the outside.

Fire-brick Lining.—A fire-brick lining was built up through the entire height of the chimney, commencing at junction of flues in foundation with a thickness of 18″, and finishing at top 5″ thick.

The internal diameter, when finished with lining, is 11′ and constant throughout its height.

The radiated fire-bricks were of five sizes, purposely made.

Stability.—The chimneys built on this plan are calculated to withstand 50-lb. wind pressure per square foot with safety. The constructors say the climate of America is dry and no doubt better for such structures than the climate of England. They believe that no one alive at the present time will see the end of a W. I. chimney lined with brick. The oldest ones in America show no material deterioration.

Painting.—The wrought-iron chimneys in America are painted every three or four years with oxide of iron paint, preferably anhydrous.

Cost.—Complete, 13,000 dollars; or £2,708. 6s. 8d.

SAMUEL FOX & CO., STOCKBRIDGE WORKS CHIMNEY, DEEPCAR, NEAR SHEFFIELD.

Builder . . . . MR. MATTHEW BREARLEY.

Description.—Octagonal brick shaft, erected in 1866: six months occupied.

Dimensions,—

Height, including foundation	201'	0"
,, from ground line to top	186'	0'
Outside measurement over sides at foundation	19'	3"
Inside ,, ,, ,, ...	5'	9"
Outside ,, ,, at ground line	14'	9"
Inside ,, ,, ,, ...	5'	9"
Outside ,, ,, at top	8'	1"
Inside ,, ,, ,, ...	5'	9"

Foundation.—Ashlar stone and brick.

Inner Shaft.—A fire-brick lining is provided for 90' high in 9" work, with a 3" cavity between lining and shaft proper.

Bricks.—250 *m.* used.

Weight.—Total, 1,000 tons.

Scaffold.—Cost about £20.

Batter.—¾" to 1 yard, or 1 in 48.

Lightning Conductor.—Copper wire.

BURY CORPORATION CHIMNEY.

Engineer, J. CARTWRIGHT, Borough Surveyor; *Builder,* CAMMICK DENNIS, Bury.

Description.—Circular brick.

Dimensions,—

Height, including foundations	195′	9″
,, from ground line	180′	0″
Outside diameter at ground line	17′	4″
Inside ,, ,,	11′	4″
Outside ,, at top	8′	4″
Inside ,, ,,	6′	0″
Height of inner brick shaft	30′	0″
Inside diameter of ,,	6′	0″

Foundation.—The foundation bed is loamy clay, on which is laid a bed of concrete 32′ square and 6′ 9″ high, the footings are stepped from 30′ square to 17′ 4″ and have a height of 9′.

Inner Shaft.—An inner shaft 6′ diameter is built parallel for a height of 30′, the brickwork is 9″ in thickness, the inner 4½″ of which is fire-brick. Between the outside of the inner shaft and the inside of the chimney proper there is an annular space of 1′ 11″ at ground line.

Thicknesses.—The chimney is divided into four sections, each of 45′ in height and having the respective thicknesses as follows, beginning at the ground line, 2′ 3″, 2′, 18″, 14″.

The bottom section is stepped out externally to 3′ in four set-offs, the highest being 1′ above ground line.

Cap.—The moulded cap is of fire-clay.

Construction.—The scaffold was an inside one. Purposely made radiating bricks were used. The shaft has a regular batter of ¾″ to the yard, and was constructed in 1881, during the months of March to September inclusive.

Lightning Conductor.—Copper rod 213′ long, cost of which fixed was £21. 6s.

Cost.—Complete, £750.

BRADFORD CORPORATION CHIMNEY, HAMERTON STREET.

Engineer, J. H. COX, Borough Surveyor; *Builders,* Messrs. NAYLOR & SMITH, Bradford.

Description.—Circular brick shaft, built 1880, in connection with "Fryer's Destructor" for town refuse.

Dimensions,—

Total height, including foundations	192′	0″
Height from ground line to top	180′	0″
Concrete foundation square	30′	0″
Outside diameter at ground surface	14′	1″
Inside „ „ „	9′	7″
Outside „ at top	9′	0″
Inside „ „	6′	8″

Thicknesses.—The shaft is composed of four sections, the lowest being 2′ 9″ in thickness and the highest 1′ 2″.

Fire-brick Lining.—A fire-brick lining 8′ internal diameter is built to a height of 45′, apart from the shaft proper, thus leaving a cavity between.

Construction.—The foundation rests upon tough clay of a somewhat blue colour, known locally as "Bowling Tough." The time occupied in building was nearly five months. Inside scaffold was used. The batter of shaft is ¼″ to a yard, or 1 in 72. The proportion of diameter to height above ground is as 1 to 12·7.

Weight.—480 tons.

Cost.—£567.

CHIMNEYS AT GEORGS MARIEN IRON WORKS, NEAR OSNABRÜCK.

Fig. 39.

1st Chimney.—*Dimensions.*—The first chimney, erected in 1857, has a height of 180′.

Clear width at the bottom of 10′ 9″
     „   „   top   „   7′ 0″

Height of square basement built in sandstone 30′.

Octagonal shaft of bricks, the walls of which are constructed of six different thicknesses, consisting of 1½ bricks at the top, and of 4 bricks at the bottom. The bricks are 10″ long.

2nd and 3rd Chimneys.—Two other chimneys, 102′ high, have also a square basement of sandstone, and an octagonal shaft of bricks, with a clear width of 4′ at the top and of 6′ 3″ at the bottom; the walls of these shafts are each 20″ thick at the top and 30″ thick at the bottom.

4th Chimney.—Built in four segments. A fourth chimney, erected in 1868, has a height of 102′, a clear width of 7′ at the top and 9' at the bottom. The shaft is built in four thicknesses of 1, 1½, 2 and 2½ bricks, and the base is 18' in height, 3' 6" thick and built of sandstone.

The shaft of this chimney is built from the base to the top in such a manner that it forms in the circumference four separate parts, which are simply brought in contact with each other.

This arrangement is shown in diagram No. 39, in which the vertical joints are designated by letters *a, b, c, d.*

This construction has been found to answer exceedingly well, and is recommended for chimneys exposed to various temperatures, as, for example, in the case of chimneys connected with coke ovens, the gases of which are either applied to the firing of boilers, or are allowed to escape directly into the chimney in the event of the boilers being laid off for cleaning or repairs.

The latter arrangements have, however, here been supplemented by the admission of cold air into the chimney together with the hot gases.

5th Shaft.—This chimney was built in 1870, and is sufficient for twenty Cornish boilers, and has a height of 120' and a circular shaft of a clear diameter of 10' at the top, and 10' 10" at the bottom.

The thickness of the walls decreases in seven steps from four bricks at the base to one brick (10") at the top.

DENS WORKS, DUNDEE, MESSRS. BAXTER BROTHERS.

Figs. 40, 41, 42 *and* 43.   *No.* 3 *Chimneys.*

Description.—Square brick pedestals; square taper brick shafts with pyramidal tops.

Dimensions,—

		Chimney No. 1.	Chimney No. 2.	Chimney No. 3.
Total height, including foundations		193' 0"	—	112' 6"
Height from ground line to top		174' 0"	135' 0"	102' 8"
,, ,, to base of pyramidal top		162' 0"	126' 0"	97' 11"
Height of square brick pedestal		34' 0"	25' 0"	25' 11"
Foundation below ground		19' 0"	—	9' 10"
Outside measurement at foundation		21' 0"	—	—
Inside ,, ,, ...		9' 6"	—	4' 3"
Outside ,, at ground line		16' 3"	13' 3"	8' 9"
Inside ,, ,, ...		9' 6"	7' 0"	4' 3"
Outside ,, at base of pyramidal top .		7' 6"	5' 9"	3' 0"
Inside ,, ,, ,, .		6' 0"	4' 3"	1' 6"
Erected		1854	1844	1864

Boilers and Flues, No. 1 Chimney.—This shaft carries away the products of combustion from No. 19 boilers, each with two furnaces, as follows :—

15 boilers 63' below the base of chimney = 225' from firing level to base of taper top
4 „   86'   „   „   „   = 248'   „   „   „

The smoke and gases are conveyed from both of these ranges of boilers by a long sloping brick flue or tunnel mostly under ground.

When chimney No. 1 was designed it was of large area for the twelve boilers it was intended to serve, since then seven boilers have been added, making a total of nineteen. When it had only a few boilers connected with it soot collected inside to a considerable extent, and occasionally caught fire and burned out in sparks and showers of smut. As boilers were added the chimney became free from soot, and the inside surface of bricks clean.

Mr. Peter Carmichael, in a paper read before the Institute of Engineers in Scotland, stated—" In our practice invariably as more boilers and furnaces have been added to a chimney the draught has been improved ; and it is obvious that if the opening in the chimney be too large compared with the whole of the openings at the dampers passing into it, the draught will be reduced."

Taper Tops.—It will be seen from diagrams Nos. 42 and 43, illustrating shaft No. 1, that the tops are constructed in the form of a pyramid, by cross walls being built from each of the four corners ; the advantage of this arrangement is given by Mr. Carmichael as follows :—" The taper top is found to answer the purpose well, the smoke ascending from it very freely, especially when there is a breeze of wind. At such times the ordinary top is acted on like a key when blown into to make it whistle, the blasts of wind affecting very perceptibly the draught of the furnaces. In the taper top this is not much felt as the wind can only blow into one or two of the four compartments at a time, and this still allows the other two to vent freely."

In the discussion that followed Mr. Carmichael's paper the taper top, which *diminishes* the area at outlet, was opposed by several members, who advocated *increasing* the area at top.

Oscillations.—At Dundee on Saturday, the 13th day of Feb., 1864, there was a violent storm of wind, which was at its height between 2 and 3 p.m. The tops of shafts Nos. 1 and 2 were blown down without injury to the shafts. The storm being a more violent one than any that had occurred for twenty years, Mr. Carmichael was anxious to see how the chimneys stood the gale ; the movement of No. 1 shaft was plainly visible, and was a steady rocking motion like the swing of a pendulum. The oscillation did not appear to exceed 12", and the observations gave a feeling of security as to the stability of the chimney.

The taper tops were re-built with bricks much heavier than those blown down, and were modelled so as to dovetail together, as shown in Figure 42. While the tops were off there was not much difference in the draughts, the cross-walls at top not being injured ; but there was more dark smoke than before, and after, and it did not rise so freely, but in a breeze of wind fell down the leeward side and clung more to the chimney.

These chimneys are still working satisfactorily, and have withstood several severe gales. The owners consider the taper construction an advantage, and the shafts "yield like a fishing rod."

Consumption of Coal.—Mr. Carmichael, in his paper, laid before the Society some valuable data with respect to the coal consumed, draught and areas of chimneys, which we have tabulated as follows :—

No. of Boilers to each shaft.	Area of shaft.		Contract'd at outlet by cross walls to sq. ft.	Area of shaft for each Boiler.		Draught recorded in inches.			Coals consumed per week of 60 hours, in tons.	
	Bott. sq. ft.	Top. sq. ft.		Bott. sq. ft.	Top. ft.	Highest	Aver.	Lowest.	Each Boiler.	Total Boilers.
1	19	90·25	36	25	4·75	1.31	·88	·8	·55	11	210
2	7	49	18·06	13·78	7	1·96	·875	·75	·6	11	75
3	1	18·06	2·25	1·75	18·06	1·75	·537	·5	·45	10	10

For twelve months he had the coals and water supplied to two ranges of boilers, one of four and the other of seven in a range, recorded weekly, with the following results :—

Range of Four Boilers.—Two flues to each boiler, area of opening at fire bridge 123 square inches coals consumed 45 tons per week of 60 hours, or 15-cwt. per hour.

$$\therefore 123 \times 8 = 984 \text{ square inches total opening} \div \text{ by coals}$$

$$\text{per hour} = \frac{984}{15} = 65 \text{ square inches for 1-cwt. per hour.}$$

Range of Seven Boilers.—Two flues to each boiler, area of opening at fire bridge 123 square inches coals consumed 75 tons per week of 60 hours, or 25-cwt. per hour.

$$\therefore 123 \times 14 = 1722 \text{ square inches total opening} \div \text{ by coals}$$

$$\text{per hour} = \frac{1722}{25} = 68 \text{ square inches for 1-cwt. per hour.}$$

Each boiler had two furnaces, each furnace was 2' 9" wide, and had two lengths of bars each 3½' long, and consumed 11 tons in 60 hours.

$$\therefore 2·75 \times 3·5 \times 2 = 19·25 \text{ square feet} = \text{fire grate area of one furnace}$$

$$\therefore 38·5 \text{ square feet} = \text{————————————} \quad ,, \quad ,, \quad \text{to one boiler.}$$

$$\therefore \frac{11 \text{ tons, or } 24,640\text{-lbs.}}{60 \text{ hours} \times 38·5} = 10·6\text{-lbs. per square foot per hour.}$$

The 123 square inches area of opening at fire bridge was adopted, after careful experiments. Mr. Carmichael stated in his paper he was of opinion that the narrow throat produces a higher temperature in the furnace, and the gases are more perfectly mixed and consumed while passing through the narrow opening, and said that at such a high temperature there was no discharge of small unconsumed cinders from the chimney.

Temperature.—The temperature of escaping gases is obtained by using small strips of the following metals:—

Zinc, which melts at 736° Fah.
Lead, ,,   ,,   612° ,,
Bismuth ,,   ,,   495° ,,
Tin, ,,   ,,   442° ,,

Small bits of each of these, about 1″ long × ¼″ broad, are pierced with a hole for suspension by a wire in the flue behind the damper, or at the bottom of the chimney, and the time occupied in melting recorded.

Mr. Carmichael, from repeated observations under various circumstances, found that the temperature of the escaping products of these chimneys was uniformly 600° behind the dampers; tin melted at once, bismuth generally in less than a minute; lead melted when the fires were in good condition, and zinc did not melt.

CROSSNESS, METROPOLITAN BOARD OF WORKS.

Engineer . . . . . Sir JOSEPH BAZALGETTE.
Built—Commenced May, 1863; Completed January, 1865.

Description.—Square brick shaft with curved base on pedestal, with stone mouldings, surmounted by ornamental iron cap.

Dimensions,—

Height, including foundations, to summit of iron cap . . . . 246′ 7″
,, from ground line to top of stone moulding under iron cap . . 177′ 0″
Foundation bed below ground line . . . . 40′ 0″
Outside dimensions, concrete foundation . . . . 30′ 2″ × 28′ 6″
,, measurement at ground line square . . . . 26′ 6″
,, ,, parallel shaft square . . . . 12′ 0″
Inside diameter   ,, flue circumference . . . . 8′ 3″

Foundation.—Foundation bed is 24′ below ordnance datum. The concrete block is 16′ in height, and upon it are laid two courses of 6″ York landings before commencing footings to shaft.

Fire-brick Lining.—The shaft is lined with Stourbridge fire-bricks for a height of 40'.

Materials.—Cement concrete (6 to 1 and 8 to 1), 800 cubic yards; bricks, about 400,000.

Men.—The bricklayers employed varied, but when above the curved portion 4 were engaged on the shaft.

Progress.—Varied very much, but the general run was 10' per week.

Scaffolding—Outside, cost about £600.

Cost.—Approximate, £4,000.

WOOLWICH ARSENAL, GUN FACTORY CHIMNEY.

Description.—Square pedestal, octagonal shaft.

Dimensions,—

Height from ground line to top	170'	0"
,, of pedestal above ground	30'	0"
Outside measurement of base at ground surface . . .	18'	9"
,, ,, bottom of shaft	13'	1"
Inside ,, ,, ,, ...	9'	4"
Outside ,, at top	5'	6"
Inside ,, ,, ...	4'	0"

Construction.—The chimney is built up in 5 sections. The heights and thickness are as follows :—

1st section . . .	pedestal	30' high	3 bricks thick.
2nd ,, . . .	shaft .	50' ,,	2½ ,,
3rd ,, . . .	,, .	30' ,,	2 ,,
4th ,, . . .	,, .	30' ,,	1½ ,,
5th ,, . . .	,, .	30' ,,	1 ,,

170 ,, above ground.

Cap.—Stone, supplied by S. Trickett, Millwall, London.

MASSACHUSETTS, U.S.A., AMERICAN PRINT WORKS, FALLS RIVER, CHIMNEY SHAFT AND VENTILATOR.

Figs. 44, 45, 46 and 47.

Engineer . . . J. A. MILLER, New York.

Height from ground line to top	160'	0"
Chimney proper internal diameter throughout	6'	0"

Hot Chamber.—It will be seen upon reference to diagram No. 46 that at the base of the shaft an enlargement or "hot chamber," 20' high × 10' diameter is formed, into which the gases delivered by the flues are discharged. The object of this chamber is to allow the currents from the flues to be gradually diverted upwards, and is of more advantage when two or more flues discharge into the same chimney, and where there are consequently conflicting currents. This chamber is vertically of eliptical form, being contracted from 10' diameter at centre to 5' diameter at outlet.

Chimney Proper.—The circular chimney proper commences from the top of the hot chamber, and is internally 6' diameter throughout its entire height. The brickwork is increased in thickness downwards by external offsets, as see diagram No. 46, so as to avoid internal projections, which, in the engineer's opinion, cause eddies, and obstruct the upward flow of the gases.

Outer Ventilating Shaft.—The outer shaft is octagonal upon a square base, about 20' high. Four of the eight sides project for the width of a brick beyond the line of the others (see Figure 47). This projection, it is said, besides improving the architectural effect of the shaft, enables the work to be executed with fewer cut bricks than are ordinarily required in octagonal or circular shafts. Between the exterior of the inner shaft and the interior of the outer one there is an annular flue of an area of 2,200 square inches, which is used for ventilation purposes, being specially suitable by reason of the heated state of the air in it, caused by the high temperature of the gases in the inner shaft.

Cap (Fig. 45).—The inner shaft is terminated a little below the top of the outer one, and by piercing the latter with openings the wind passes through and impinges upon the sides of the cap of the inner shaft, the shape of which causes it (the wind) to be deflected upwards, and thus, by the action of induced currents, assists the draught rather than diminishes it, as is the case with the ordinary top, where it acts to a certain extent as a damper and checks the draught.

The cast iron caps to both the inner and outer shafts are built up in sections, each of which is of such shape that its centre of gravity falls outside the inner line of the bases of the sections, and the sections thus tend to fall together and form a kind of arch. Each section is connected to that adjoining it by two bolts, but from the shape given to the sections they will maintain their position, even if the bolts are corroded away. The sections are provided at their bases with flanges, which grip the shaft both within and without, as see Fig. 45.

MR. W. D. BARKER, PATENT BRICK WORKS, WORCESTER.

Architect, F. CHAMBERLAIN, Barnsley; *Builder,* W. D. BARKER.
Built in 1869, from March to July (about four months).

Description.—Circular brick chimney, outer and inner shafts.

Dimensions,—

Height, including foundations		164' 0"
„ from ground surface to top		160' 0"
Diameter of circular brick foundation		24' 0"
Outside diameter at top of footings		13' 10"
Inside „ „ „		7' 7½"
Outside „ „ under cap		6' 9"
Extreme „ of cap		10' 6"
Inside „ „		4' 6"

Foundation.—The foundation bed is rock marl, 4' below ground surface, upon which the brick footings commence, and are carried up to ground level.

Construction.—The *Outer Shaft* from top of footings to a height of 22' is of 14" work, and from thence for a height of 110' (28' from top) is 9" in thickness. The alteration from 14" to 9" work is made by an external set back of half a brick, thus giving the appearance of a pedestal.

The *Inner Shaft* for a height of 10' from footings is in 9" work, thence to within 28' from top is 4½" in thickness; at this height the *outer* and *inner* shafts merge into one, and are continued in 14" solid work to under cap. The inner shaft is every 10' in height bonded to the outer one by No. 12 radial brick ties.

The shaft was built by day labour, under the supervision of Mr. Barker.

Brickwork.—92,000 purposely made red bricks 9" × 4½" × 3¼" were used in the construction of this shaft, costing 26s. per *m.* The whole of the bricks in the exterior shaft were made to suit the radius of the stack. The lime and sand were ground together in a mill, and used almost hot.

Bond.—Three stretchers to one header.

Weight.—390 tons = ·86 ton pressure per square foot on foundation bed.

Batter.—The batter of the *outer* shaft is ¾" per yard, that of the *inner* shaft is ¾" per yard.

Scaffold.—Inside.

Duty.—One boiler and one Hoffman's brick kiln.

Cap.—The cap is of brick formed by a course of round end bricks 18″ × 6″ × 4″, then two courses of white bricks cornered out, then two courses of ordinary red bricks, then a course of round ends 18″ × 9½″ × 6″, then gathered in and finished with half round coping 14″ × 7″. This work is in cement, and the large bricks were cut to the radius of the cap. Mr. Barker says this cap gives a pleasing finish to the shaft, and is excellent in its simplicity, and he cannot understand why heavy cast iron caps are adopted when a better effect can be obtained in brickwork.

Lightning Conductor.—Copper, costing about £11 fixed.

WROUGHT IRON CHIMNEY, MESSRS. FRANCIS & CO., THE NINE ELMS CEMENT WORKS, CLIFFE CREEK, ROCHESTER.

Erected, 1878; *Designed* by Mr. V. DE MICHELLE, C.E.; *Constructed* by Messrs. FIELDING & PLATT, Gloucester.

Description.—The shaft is circular and parallel throughout, and is constructed as follows :—

Dimensions,—

Height from ground line to top . . . . 160′ 0″
External diameter throughout . . . . 5′ 0″
Internal   „   „   . . . . 4′ 6″

Wrought Iron Plates.—The plates vary in thickness downwards, from ¼″ to ⅜″.

Fire-brick.—The shaft is lined with 3″ fire-brick its entire height.

Wind Stays.—The chimney is stayed against the wind by No. 4, 3¾″ steel guy ropes.

Duty.—This chimney was erected over the centre one in a row of nine cement kilns, which are all connected to shaft by a wrought iron horizontal flue 4′ in diameter. Two additional kilns have since been added, and the chimney now carries off the gases from eleven cement kilns.

Chimney Base.—Round the outside of centre kiln on ground level is fixed a cast iron curb or base plate. On this base stand four cast iron standards or supports, having their lower ends butting into and secured to base plate. These standards incline inwards until their upper ends meet to support a cast iron circular chimney base, which forms the top of the centre kiln. The wrought iron chimney proper commences from top of this circular cast iron base, directly over which the 4′ horizontal flue is connected to shaft.

Construction (Novel Erection).—A timber stage was erected at about level of kiln tops, and upon this stood the rivet fires. Four winches were worked on this stage, and to them were led guy ropes, after passing round blocks at convenient distances. A hydraulic press with a 4′ stroke was then fixed over the centre kiln, and the top length of 20′, which had previously been riveted up on the ground and raised to the stage level, was placed upon the ram. The ram was then pumped up, and the 20′ length raised a height of 4′, the guy ropes being slackened out to the required extent, as the 20′ length gradually rose. A 4′ ring of plating was then riveted on with $\frac{3}{4}''$ snap head rivets, and the usual lap. The ram was then again pumped up, and the now 24′ length raised the necessary height; another ring of plates was then riveted on, and the operation repeated until the chimney had reached its required altitude.

The Engineer says "The chimney answers its purpose admirably." On a bad foundation he would recommend the construction of an iron shaft, but on a good one, decidedly brick.

Cost.—About £1,000, including long wrought iron flues.

COTTON FACTORY, MEXICO.

Description.—Chimney built of apparently sun-dried bricks, in use over twelve years, and in excellent condition. This shaft was built by Indians, is symmetrical, and appears to be well constructed.

Height about . . . . 160′ 0″
Bricks, size of . . . . 10″ × 3″ × 7″

The above particulars were given at a meeting of the American Society of Civil Engineers, March 5th, 1884.

At the same meeting a chimney was mentioned as being in successful use in Pennsylvania, U.S.A., which is generally known as the "crinoline chimney," being constructed of old rails.

G

MR. SHEKOLDEN'S PAPER MILL, ADESHAVO (NEAR THE TOWN
OF KINESHMA), KOSTROMA GOVERNMENT, RUSSIA.

Engineer . . . . . V. T. GREGORY, Moscow.

Description.—Circular wrought iron shaft, brick pedestal.
Erected March, 1874.

Dimensions,— .

Total height, including foundation	170'	0"
Height of brick pedestal above ground	14'	0"
Height of W.I. shaft 62 Archines = ...	144'	8"
Total height from ground to top	158'	8"
Diam. inside, at foundation, pedestal, and top of shaft 2 Archines =	4'	8"
Outside measurement at brick foundation footings	11'	8"
Pedestal outside measurement at ground surface	9'	4"
„ „ „ at top	7'	0"

Thickness.—The shaft is circular and parallel all the way
up, and composed of plates varying in thickness downwards
from $\frac{3}{16}$" to $\frac{3}{8}$".

Duty.—This chimney carries off the gases from the flues of
five boilers, which give in the aggregate 250 nominal h.p.
Wood fuel is used, and the chimney draught travels some
distance and passes round a large wrought iron tube placed in
flue as a water heater to feed boilers, consequently the heat at
bottom of shaft is not so intense as to require any fire-brick
lining to protect the shaft.

Foundation.—It was the intention of the owner to build a
brick chimney, but finding the ground very bad he decided to
have one in wrought iron. The foundation is of brickwork on
gravel, and is carried to 11' 4" below ground level.

Construction.—The chimney was made at contractor's works,
and brought to the mill in three parts and riveted together near
site. The sizes of rivets used were $\frac{5}{8}$" diameter in lower portion
and $\frac{3}{8}$" diameter at top, the lap of plates being about 2$\frac{1}{4}$".

The appliances used for lifting were a pair of sheer legs
and two pairs of blocks. The legs were built on ground with
pine timber cut in the adjoining woods, and bound together
with wrought iron hoops and bolts ; then taken to pieces after
each timber had been marked to its own place. The legs were
then removed to the chimney site and erected so as to lean over
centre of foundation. The blocks were then hooked to, and the
tube lifted in bulk and placed on its brick pedestal.

The erection was carried out by Mr. J. P. Moorhouse, of Burnley, England, and the time occupied in lifting, after all preparations had been made, was 1½ hours.

Mr. Moorhouse states "wrought iron chimneys are very common in Russia, and seem to give general satisfaction. They are used principally for small or medium sized works of all sorts, the first cost being much in their favour."

The Russian climate being much drier is more favourable to wrought iron shafts than the English.

ABBEY MILLS PUMPING STATION, METROPOLITAN BOARD OF WORKS.

Engineer, SIR JOSEPH BAZALGETTE. *Built,* 1867; time occupied in building brick shaft from top of stone base to underside of stone head, 20th July, 1867, to 9th Nov., 1867.

Description.—Octagonal brick shaft with curved base on massive square stone moulded pedestal; top formed of stone, surmounted by large ornamental iron cap.

Dimensions.—Taken from contract drawing,—

Height, including foundations to summit of iron cap . .	212'	0"
Foundation below ground line	21'	0"
Height from ground line to top of stone cap	158'	3"
Height of pedestal above ground line	30'	9"
Outside measurement, square concrete foundation . . .	43'	0"
,, ,, square brick footings	39'	6"
,, ,, at ground line	30'	6"
,, ,, top of pedestal	17'	6"
,, ,, top of shaft under cap	10'	3"
Inside ,, throughout, except at set offs . .	8'	0"

Foundation.—The foundation bed is 6' below Ordnance Datum. The concrete layer is 6' deep and 37' 6" × 37' 6"; it is, however, carried out on each side 2' 9" to form foundation to base buttresses. Upon the concrete the footings are constructed for a height of 4', having four steps and set offs.

Shaft.—From pedestal to underside of cap is constructed in three sections, as follows :—

1st section, including curved base 40' high tapering from 4' 4½" to 1' 10¼" thick.
2nd ,, ... 39' ,, ... 1' 6" ,,
3rd ,, , ... 39' ,, ... 1' 1½" ,,

Brickwork.—From stone base to underside of stone head, time,—

Bricklayers . . . . . 525 days.
Labourers . . . . . 590 ,,

This equals a cost of £8 per rod for labour only.

Portland Stone.—Stone cap, 900 cubic feet or 60 tons, supplied by S. Trickett, Millwall.

Time for raising and fixing only—
Masons . . . . . . 72 days.
Labourers . . . . . 144 ,,
This equals about 1s. 2d. per cubic foot.

Scaffolding, &c.—The means used for raising the bricks and stone on shaft were by a small donkey engine, chains being fixed round wheels at top and bottom of shaft, after reaching 50′ in height.

No. of scaffold poles used—
Shaft 428
Barrow lift. 18
                      ———
                      446 total.

No. of cords used—
Standards 324
Barrow lift. 54
Braces 216
Ledgers 544
                      ————
                      1138 total cords.

Iron Cap.—Weight, 27 tons. This cap was ultimately removed during the examination and repair of top of shaft in November, 1883, owing to the chimney being very much fissured by cracks. The masonry at top was strengthened by three additional braces of iron.

SALFORD CORPORATION SEWAGE WORKS.

Fig. 74.

Engineer, ARTHUR JACOBS, M.I.C.E.; *Builders,* S. W. PILLING & CO.; *Built,* 1883.

Description.—Octagonal brick shaft, on square ornamented pedestal.

Dimensions,—

Total height, including foundations . . . . 180′ 6″
Height from ground line to top . . . . 156′ 6″
   ,,   of pedestal square and parallel . . . . . . 40′ 0″
Outside measurement at ground line . . . . 17′ 6″
Inside   ,,   ,, ... 5′ 3″
Outside   ,,   base of octagonal shaft . . . . 11′ 6″
Inside   ,,   ,,   ,, ... 6′ 0″
Outside   ,,   at top . . . . 8′ 4″
Inside   ,,   ,, ... 5′ 3″

Foundation Bed.—Red sandstone.

Construction, Thicknesses, &c.—Pedestal, 40' high, five bricks thick, exclusive of fire-brick.

Shaft 1st section	. . . . . .	34'	6" high	3½	bricks.
2nd „	. . . . . .	31'	7" „	3	„
3rd „	. . . . . .	31'	7" „	2½	„
4th „	. . . . . .	18'	10" „	2	„
Height from ground line	. . . .	156'	6'		

The chimney was built slowly. The workmen constructed about 2' in height daily, and when each 6' was built the work was left to rest for a couple of days.

Ordinary bricks, laid to Flemish bond, used in construction; best pressed bricks used outside, and stock bricks on the inside. At angles purposely-made bricks were employed. No hoop iron used.

Fire-brick Lining.—This shaft has a fire-brick lining 40' high, 20' one brick thick and 20' ½ brick, 1" cavity being left between pedestal and fire-brick lining.

Scaffold.—Inside scaffold was used, consisting of No. 4 put-logs, covered with boards for platform. A small set of shear-legs above.

Duty.—The flues from No. 4 boilers are connected to shaft; the boilers are 27' long × 7' diameter, with two flues to each boiler.

MILE END SPOOL COTTON MILL, EAST NEWARK, N.J., U.S.A.

Engineers . . . BABCOCK & WILCOX COMPANY, New York.

Description.—Octagonal brick shaft, on square brick base, surmounted by ornamental cap.

Dimensions,—

Height above footings	161'	6"
„ above ground line	150'	0"
„ of square base above ground	30'	0"
Outside measurement, parallel square base	17'	0"
„ „ foot of shaft	16'	6"
„ „ at top „	12'	0"
Inside „ throughout, except at off sets . .	10'	0"

Pedestal.—The square base extends 11' 6" below and 30' above ground line. It measures outside 17' × 17' its entire height, and the walls are 32" in thickness.

Shaft.—Above the square base the shaft is gradually stepped at the corners until the shaft is octagonal in form. It is built in four sections, as follows :—

1st section, including stepping	.	30' high	24" thick.
2nd ,,	. . . .	30' ,,	20" ,,
3rd ,,	. . . .	30' ,,	16" ,,
4th ,, including cap	. . .	30' ,,	12" ,,

Inner Shaft.—An inner shaft is constructed for a height of 80' above ground line, as follows :—

1st section	. . . .	50' high	9" thick.
2nd ,,	. . . .	30' ,,	4½" ,,

It is 10' diameter inside, thus leaving a cavity at the angles between itself and the brickwork of square pedestal and octagonal shaft. A wall across the circular flue at the base of the chimney, is built diagonally for the purpose of giving a larger and easier line from the horizontal to the vertical flue. It is also intended to prevent the two currents of gases from interfering with each other until they begin to travel in the same direction.

Batter.—1 in 53·3.

Duty and Flues.—The boiler plant when complete will be composed of two sets or divisions of 1,248 h.p. each. The right hand division remains vacant for the present, the one at the left being all that will be necessary for some time to come. The entire plant will consist of six independent batteries of boilers. Each battery is composed of two independent boilers of 208 h.p. each. The flue for conveying the products of combustion from the boilers to the chimney is double, and the gases can either be thrown into a direct flue underneath the economizer to the chimney, or can be passed up through an 8' × 5' damper, thence across the economizer into the stack.

Plan.—Experience has demonstrated that it is best to divide the economizer into two parts, as it makes the flues more direct, and allows one to be in use while the other is being cleaned or repaired, as may occur in an emergency. Furthermore, this arrangement makes a more symmetrical plant.

The ground level of the boiler-house is the street level, and a 7' door in front of the boilers is built for dumping the coal from the street directly into the boiler-room.

Cap.—The cap commences 12' from the top, and is constructed for a height of 8' in brickwork, the remaining 4' being in stone, the whole forming a neat but ornamental cap.

EASTBOURNE WATER WORKS CHIMNEY.

Fig. 76.

Engineer, G. A. WALLIS, M.I.C.E.; *Architect,* H. CURREY, F.R.I.B.A.
Built, 1881-2; 5 months occupied in building.

Description.—Square brick shaft, buttressed.

Dimensions,—

Total height, including foundation . . . . 165′ 0″
Height from ground line to top . . . . 150′ 0″
Concrete foundation square . . . . 30′ 0″
,,   ,,   deep . . . . 8′ 0″
Outside measurement at ground surface . . . . . . 12′ 0″
Inside   ,,   ,,   ,, ... 6′ 0″
Outside   ,,   at top . . . . 6′ 0″
Inside   ,,   ,, ... 3′ 0″

Foundation Bed.—Chalk.

Buttresses.—This shaft has two buttresses on each side, making a total of eight; they are 3′ in thickness, and extend to a height of 34′.

Batter.—1 in 50.

Materials.—134,850 bricks were used in the construction of this chimney.

Weight.—542½ tons.

Scaffold.—Outside scaffolding was used at a cost of £101. 5s.

Lightning Conductor.—Copper tape.

MESSRS. PROCKTER & BEVINGTON, GLUE AND SIZE WORKS, BERMONDSEY.

Architects, GEO. ELKINGTON & SON; *Builder,* BENJAMIN WELLS.
Built, 1881-2; from June to January, 8 months.

Description.—Circular.

Dimensions,—

Total height, including foundations . . . . 161′ 6″
Height from ground line to top . . . . 150′ 0″
Outside measurement of square foundation . . . . . 24′ 0″
Inside   ,,   ,,   ,, ... 6′ 0″
Outside   ,,   at ground surface . . . . . . 16′ 0″
Inside   ,,   ,,   ,, ... 6′ 0″
Outside   ,,   at top . . . . 6′ 0″
Inside   ,,   ,, ... 4′ 6″

Construction.—The bond of brickwork is principally English. The shaft contains 38 rods reduced brickwork. Ordinary best stock bricks were used.

Scaffold. — Outside scaffolding was used for first 100′, then inside scaffold to top.

Lightning Conductor. — The conductor is of ¾″ copper stranded rope.

LONDON AND NORTH-WESTERN RAILWAY, BROAD STREET GOODS STATION, LONDON.

Fig. 75.

Builder . . . . JOHN JAY.

Description.—Octagonal brick, erected 1867, for boilers.

Dimensions,—

Height, including brick footings	170′	7″
„ from ground line to top	150′	0″
„ of footings	14′	0″
Base „ . . . . 36′ 0″ × 22′	1¼″	
Outside measurement over sides at ground line	15′	0″
Inside „ of fire-brick shaft	4′	9″
Height of 9″ fire-brick shaft from base	55′	3″
Outside measurement under cap at top	8′	0″
„ „ at top of shaft	7′	6″
Inside „ „ „	4′	6″

Thicknesses. — The octagonal shaft is built up in eight sections, from top of footings 6′ 7″ below ground, as follows :—

1st section	9′ 6″ high	4′ 4¼″ thick.		
2nd „	16′ 0″	„	3′ 11″	„
3rd „	16′ 0″	„	3′ 6½″	„
4th „	21′ 6″	„	3′ 1½″	„
5th „	21′ 6″	„	2′ 9″	„
6th „	27′ 0″	„	2′ 3½″	„
7th „	22′ 6″	„	1′ 11″	„
8th „	22′ 7″	„	1′ 6″	„

Total height above footings 156′ 7″

Batter.—1 in 40.

MESSRS. E. BROOKS & SONS, FIELDHOUSE FIRE-CLAY WORKS,
HUDDERSFIELD.

Figs. 48 *to* 53 *inclusive.*

Architect, B. STOCKS, Huddersfield; *Builders,* E. BROOKS & SONS. *Built,* 1878.

Description.—Circular brick chimney, outer and inner shafts.

Dimensions,—

Height, including foundations	158'	3"
„ from flue invert to top	150'	0"
Diameter of circular concrete foundation	24'	4"
Outside diameter top of footings	16'	8"
Inside „ „ „	7'	0"
Outside „ at top	9'	4"
Inside „ „	7'	0"

Foundation.—A bed of concrete 24' 4" in diameter forms the base for foundation footings. Upon this are laid eight stones, 12" thick and each 8' across, leaving a 12" margin of concrete, and a circular space at centre 6' 4" in diameter, which is filled in with brickwork. Upon this stone base the brickwork footings commence 21' 2" in diameter, and are stepped in nine off sets, each 6" in height, terminating 16' 8" in diameter at top. See Figs. 48 and 49.

Flues.—There are four flue openings at base of shaft, 7' high × 2' 6" wide, with 9" fire-brick sides and inverts, and 14" fire-brick top arches.

Outer Shaft.—The shaft between the flue openings is built up solid. From level of top of 14" arches to flues the outer shaft commences, with heights and thicknesses as follows:—

1st section	36' high	18" thick.
2nd „	45' „	14" „
3rd „	27' „	9" „

At this height it joins the inner shaft, and the work is continued solid to top, 34' high, and terminating 14" thick.

Inner Shaft.—This commences at level of top of flue arches—

58' 10" high	14" thick.
48' 0" „	9" „

At this height it joins the outer shaft, as above-mentioned.

The inner shaft for a height of 58' 10" is lined with firebrick, bonded into brickwork every fourth course. The top 48' is not lined.

Cavity. — The cavity between inner and outer shafts is divided by eight tie walls 14″ thick for height of 14″ work of outer shaft, and 9″ thick to junction of shafts. Each of the eight spaces thus formed is provided with means of ventilation by one opening at base and one at top 12″ square, each opening having a cast iron grating built into brickwork. See Figs. 49, 51 and 53.

Batter.—External ⅞″ per yard or 1 in 41·1; internal flue, parallel.

Cap.—A fire-brick course 14″ thick × 12″ high forms the only finish to shaft.

It was at first intended to construct a 2′ string course 4·6 from top to add to the appearance of shaft, but even this Messrs. Brooks omitted being of opinion that the plainer the termination the better.

MESSRS. E. LACON & SONS, BREWERS, GREAT YARMOUTH.

Engineer and Builder . . . Mr. WAKEFIELD, London.

Built 1838; three months occupied in building.

Description.—Square brick panelled pedestal with two 4½″ set-offs for plinths.

Dimensions,—

```
Original height, including foundation . . . . . . . 160′  0″
      „        „     from ground line to top . . . . . . 140′  0″
Present     „     „      „      „      „   . . . . . . 100′  0″
Outside measurement at ground line, square . . . . . 10′  6″
```

Scaffold.—Inside scaffolding used in erection.

Remarks.—When completed this shaft was 140′ high from ground line, but about 23 years ago it became so bent by strong north winds that the proprietors had 40′ taken off, and it now stands 100′ high.

MESSRS. W. MACFARLANE & CO., SARACEN FOUNDRY, GLASGOW.

NO. 2 SHAFTS.

Engineer, JAS. BOUCHER ; *Builders,* ROBERT CORBET & SON.

	1	2
Situation	Possil Park.	Washington Street.
Description	Round.	Square.
Height, including foundation	140½'	110¼'
„ above „	134'	105'
„ from ground line	130'	—
Outside measurement at foundation . .	9'	11½'
Inside „ „ . .	6'	7¼'
Outside „ at ground surface	9'	—
Inside „ „ „	6'	—
Outside „ at top	9'	8'
Inside „ „	5'	5¾'
Cost of scaffold	£8	£4. 10s.
Description of do.	Inside.	Inside.
Cost complete	£432	£277
Built	1873	1862
Foundation bed	Clay.	—

CARBON FERTILIZER MANURE COMPANY, OLDHAM.

Architect, A. PAYNE, F.R.I.B.A.; *Builders,* Messrs. NEILL & SONS.
Built, 1874 ; five months occupied, exclusive of foundation.

Description.—Brick, square pedestal, round shaft.

Dimensions,—

Height, including foundation	165'	0"
„ above concrete „	142'	6"
„ „ ground line	133'	0"
Outside measurement, square concrete	23'	0"
Inside „ at base	7'	0"
Outside „ square pedestal	13'	10"
„ diameter at top	7'	3"
Inside „ „	5'	0"

Foundation.—The foundation bed was very bad. A concrete block of considerable depth formed the base for footings, which were square externally, of brick, and 4' 6" in height.

Construction.—The chimney above concrete is composed of nine sections, as follows :—

		Height.
1st sec. . Footings		4′ 6″

Pedestal.

2nd ,,	. Brick in mortar, 13′ 10″ sq. externally, partly below ground	.	17′ 0″
3rd ,,	. Brick in mortar, 12′ 4″ sq. externally, with hoop iron bond	.	23′ 0″

Shaft.

4th ,,	. Moulded circular brick base to shaft in cement	4′ 0″
5th ,,	. Brick in cement, bonded with hoop iron, 9′ 6″ external diam. .	17′ 0″
6th ,,	. Brick in mortar 9′ diam., hoop iron bond at intervals. . . .	22′ 0″
7th ,,	. Brick in cement 8′ 9″ diam., with hoop iron bond	11′ 0″
8th ,,	. Brick in mortar 8′ 3″ diam., with hoop iron bond	22′ 0″
9th ,,	. Brick in cement, including cap, with hoop iron bond	22′ 0″

Height above concrete . . . 142′ 6″

Shaft.—The shaft is composed of two set offs expanding upwards.

Hoop Iron Bond.—The hoop iron was used in several courses close together, and several lengths in each course.

Brick in Cement.—The Architect had the cement rings constructed " to give greater strength to the shaft, and to obviate cracks, cement having greater adhesive power than ordinary mortar."

Fire-brick Lining.—The shaft is partly lined with fire-brick, but the heat is not expected to be excessive, it being mostly used for hypocaust drying rooms.

Cap.—The cap has a considerable projection, and is carried by a stout ring of York stones well cramped together, resting on brick corbels.

Lightning Conductor.—Copper tape, insulated.

MESSRS. GROUT & CO., SILK CRAPE WORKS, GREAT YARMOUTH.

Built, 1829-30 ; 12 months occupied in construction.

Description.—Square brick chimney.

Dimensions,—

Height from ground line to top			127′ 6″
Outside measurement at foundation			18′ 0″
,, ,, at ground surface			12′ 6″
Inside ,, ,, ,, ...			6′ 10″
Outside ,, at top			3′ 6″
Inside ,, ,, ...			2′ 0″

Foundation Bed.—Sand.

Thicknesses.—The shaft is composed of eight sections, as follows :—

Plinth from ground line . .	3′ 0″ high	3′ 6″ thick.
1st section	12′ 1″ ,,	2′ 10″ ,,
2nd ,, ...	14′ 3″ ,,	2′ 6″ ,,
3rd ,, ...	17′ 3″ ,,	2′ 2″ ,,
4th ,, ...	15′ 9″ ,,	1′ 10″ ,,
5th ,, ...	16′ 9″ ,,	1′ 6″ ,,
6th ,, ...	33′ 0″ ,,	1′ 2″ ,,
7th ,, ...	13′ 6″ ,,	9″ ,,
8th ,, cap and blocking	1′ 11″ ,,	9″ ,,

Height from ground line 127′ 6″

Scaffold.—Inside scaffolding used.

Batter.—1 in 28.

Remarks.—The brickwork to top of plinth was built in Portland cement.

On the 19th July, 1884, at 9.30 a.m., the chimney was struck by lightning, damaging the shaft to the amount of £68.

PENNSYLVANIA RAILROAD, WEST PHILADELPHIA SHOPS, U.S.A.

Figs. 54 to 59 inclusive.

Engineer . . . . . JOSEPH M. WILSON, Philadelphia.

Description.—Brick chimney, circular pedestal, shaft of an eight pointed star section.

Dimensions,—

Height from ground line	121′ 8″
,, of pedestal above ground.	7′ 8″
Outside diameter of pedestal	13′ 0″
,, ,, of shaft, over points, under cap . . .	8′ 9″
Inside diameter throughout	4′ 0″
Flue invert below ground	5′ 0″

Foundation.—Stone.

Section.—This shaft is remarkable, inasmuch as its section has the form of an eight pointed star, with a circular inner shaft, thus leaving the points of the star hollow. These hollow spaces run from 2′ above ground line to top, and have air openings in the brickwork at base and in the cast iron cap at top.

Outer Shaft and Pedestal.—The heights and thicknesses of the brickwork, starting from top of plinth, are as follows :—

 1st section, including pedestal 36' high 1' 6" thick.
 2nd „   . . . . . . 36' „ 1' 2" „
 3rd „   to 12' from top . 35¾' „   9" „

Inner Shaft.—Two feet from ground line to top :—

 1st section . . . . . 5' 3" high 1' 11" thick.
 2nd „   . . . . . 50' 0" „ 1' 6" „
 3rd, „   . . . . . 40' 0" „ 1' 2" „
 4th „   . . . . . 24' 5" „   9" „

Brick Lining.—An inner lining 4½" thick runs throughout the stack, but is not bonded to the inner shaft, for 17' from flue invert the lining is of fire-brick.

Cap (Fig. 59).—At a height of 12' from the top the brickwork is gradually stepped outward until it measures 14' in diameter; into this bell-shaped top the eight points gradually die away, and upon it a cast iron hollow cap, 3' in height, is placed, covering the brickwork of inner shaft, and having outlets for the escape of the heated air from the eight cavities before mentioned.

WOOLWICH ARSENAL.

Dimensions,—

 Height above ground . . . . 120' 6"
 Outside measurement at base . . . . 10' 9"
   „   „   at top . . . .   4' 9"

Construction.—This chimney is built up in five sections, the heights and thicknesses of which are as follows :—

 1st section. . . . . 16' 6" high 3 bricks thick.
 2nd „   . . . . . 25' 3" „ 2½   „
 3rd „   . . . . . 25' 3" „ 2   „
 4th „   . . . . . 25' 3" „ 1½   „
 5th „   . . . . . 28' 3" „ 1   „

   120' 6" high above ground.
  The top 28' 3" was built in cement.

Scaffold.—Erected independently of shaft, without putlogs.

MESSRS. REID & CO., BREWERS, CLERKENWELL ROAD, LONDON.

Engineer, GEO. SCAMMELL; *Builders,* Messrs. CUBITT & CO.

Description.—Square brick pedestal on concrete foundation about 52' high, including 9' stone moulding. Round brick shaft.

Dimensions,—

Total height, including foundation	142'	3"
Height from ground line to top	120'	0"
Outside measurement of square concrete foundation . .	22'	0"
„ „ square footings	20'	3"
Height of footings	5'	9"
Outside measurement at ground surface	12'	0"
Inside „ „ „	5'	3"
„ „ at outlet top of cap	5'	8"

Foundation Bed.—Gravel.

Construction. — The walls of pedestal are constructed in brickwork 2' 7½" thick (3⅓ bricks), then the first 21' of circular shaft 2½ bricks thick, next height of 23' in 2 bricks thick, and topmost height of 12' in 1½ bricks thick. The top of shaft is formed by a 6' length of stone work, and then terminates with a cast iron cap about 5' high, fitted together in segments, bolted internally. The outlet formed by this cap is bell-mouthed, and is 5' 8" internal diameter.

YALE LOCK MANUFACTURING COMPANY, STAMFORD, CONNECTICUT, U.S.A.

Figs. 60 to 64 *inclusive.*

Engineer, HENRY R. TOWNE. *Built,* 1881.

Description.—Square brick shaft on panelled pedestal.

Dimensions,—

Height, including foundation		126'	0"
„ from ground line		120'	0"
„ of pedestal above ground		18'	0"
„ of shaft above pedestal		102'	0"
Outside measurement at base of foundation		15'	0"
„ „ at top of „		11'	10"
„ „ at ground line		11'	10"
„ „ at top of pedestal		11'	0"
Inside „ „ „		4'	0"
Outside „ at base of shaft		10'	4"
Inside „ „ „		4'	0"
Outside „ at top under cap		6'	1"
Inside „ „		4'	9"

Foundation.—The foundation is shallow, for the reason that a bed of hard gravel was struck a few feet below the surface of the ground, which formed a good basis for the chimney foundation. The first course of stone was composed of large selected flat boulders 4′ × 4′, upon top of which were placed other smaller ones laid in Portland cement. The upper part was built of small stones grouted with cement.

Outer Shaft.—This is built up in four sections, viz. :—

1st section	. . . .	7′ high	20″ thick.
2nd „	. . . .	25′ „	16″ „
3rd „	. . . .	31′ „	12″ „
4th „	. . . .	34½′ „	8″ „

Inner Shaft (Fig. 60).—This is constructed in three sections, commencing 22′ 6″ above ground line, as follows :—

1st section	. . . . . .	9′ 9″ high	12″ thick.
2nd „	. . . . . .	36′ 3″ „	8″ „
3rd „	. . . . . .	30′ 0″ „	4″ „

The inner shaft is built independently of the outer one, its inside measurement at base being 4′ and at top 4′ 6″. From this it will be seen the inner shaft has an outward batter of ·04″ in one foot, and was so designed that it affords free escape for the products of combustion, as they expand in their progress upward.

At the summit of this inner shaft is a sheet iron hopper top, No. 18 W. G., which covers the sharp off set that would otherwise occur to affect the draught.

Iron Bond.—Tie irons, consisting of pieces of flat hoop iron, with each end bent downwards, were built into the walls all the way up, at distances of 3′ apart. The inside walls also had tie irons up to termination of 8″ work.

Iron Ladder.—An iron inside ladder extends from the bottom to the top, and is formed by building in short bars of round iron ¾″ in diameter across one of the corners of the square interior (see Figs. 63 and 64). An iron ladder is suspended from one of these bars to reach over the sheet-iron hopper top of the inner shaft (see Fig. 60).

Fire-brick.—There is a wall of fire-brick 8′ high (see Fig. 60) on the inside of chimney, where the flue enters pedestal. This wall was put up after the chimney was completed, and may be re-placed at any time without disturbing the outer walls. There is a 2″ space between the fire-brick and the brickwork of the chimney proper.

Batter.—Pedestal, parallel. Outer shaft, ¼″ in 1′ or 1 in 48.

Duty, &c.—This shaft was constructed to run six horizontal tubular boilers, 54″ in diameter and 15′ long. Each boiler has 54 tubes 3″ in diameter, with a grate 4′ long × 4′ 6″ wide, and is rated at 65 h.p. The opening from each boiler to flue is

13″ × 33″, while the main sheet iron flue is 36″ in diameter. The boilers, in addition to furnishing steam to drive a 20″ × 42″ stationary engine of the Corliss type, have to supply steam for heating extensive buildings, and for other purposes.

EXTRACTS FROM SPECIFICATION.

1. *Foundation.* — That the Company will have built and finished, ready for the brickwork, a proper stone foundation, and will assume all responsibility therefor.

2. *Bricks.*—That said Company will furnish, delivered on the ground, as the contractor may require them, all bricks required in the construction of the chimney, and that the contractor will carefully select bricks of good and uniform colour for the external faces of the chimney.

3. *Materials.*—That the contractor will furnish all other materials required, including lime and sand, all of which shall be of thoroughly good quality of their respective kinds, special care being exercised to secure a strong and durable mortar.

4. *Brickwork.*—That all portions of the chimney shall be built in strict accordance with the plans and sections shewn on the drawings attached hereto. The bricks for outside faces of walls to be selected as specified above, and no swelled, rotten, or unsound brick to be used in any part of the structure. All bricks to be well wetted before laying, and to be laid with flushed solid joints, leaving no empty spaces in the walls, except where called for on the plans. All courses to be run perfectly level and straight. The batter of outside walls to be exactly one-fourth inch to the foot, and the greatest care to be taken that a uniform batter is preserved on each of the four faces, so that the chimney, when completed, shall be exactly perpendicular. Special pains must also be taken to carry each of the outside angles exactly vertical and to avoid giving them any "twist." The joints of the exterior walls of the chimney and of the interior walls of the vertical flue (that is, those with which the smoke will come in contact) to be filled fully flush with faces of bricks when struck with the trowel, and every joint, both horizontal and vertical, to be neatly and carefully rubbed smooth with a "jointer," so as to leave the faces of the walls perfectly smooth and flush.

5. *Bonding Irons.*—That the Company will furnish proper bonding strips of hoop iron, cut to lengths required, and that the contractor will lay these strips in both the inner and outer walls

H

of the chimney, at vertical intervals of about three feet, as may be directed by the Company, the ends of each bonding iron being bent so as to engage with the brickwork and assist in tying it firmly together.

6. *Ladder.*—That the Company will furnish, and the contractor build in place, in one corner of the inside flue, iron bars, to constitute a steps or ladder for gaining access to the top of the chimney. These bars to be placed diagonally across one of the corners and at vertical intervals of about 18″.

7. *Openings.*—That the contractor will set and build into the chimney, as shewn on the drawing, an iron cleaning door (to be furnished by the Company), and will also construct two circular flue openings of the form and position shewn by the drawing.

8. *Caps.*—That the Company will furnish, delivered on the ground, a cast iron cap or funnel, to be placed on the top of the inside lining or flue, as shewn by the drawing, and also cast iron cap, in four pieces (complete with joint plates) to form the final finish on top of chimney, the contractor to raise these caps and carefully set them in their proper positions.

9. *Fire-bricks.*—That the contractor will construct a firebrick lining (the joints to be carefully laid with fire-clay) near the base of chimney, as shewn on the drawings. This fire-brick lining to be laid independently of the surrounding walls, so that it shall not carry any part of the weight of the chimney, and so that it may be removed and renewed without disturbing the surrounding walls. The fire-bricks and clay to be furnished by the Company.

10. *General.*—That the contractor will provide all necessary scaffolding and will remove the same upon completion of the contract; that the Company will make provision whereby the contractor can obtain a supply of water by a line of hose (to be furnished by the contractor), not exceeding 100′ in length; that the contractor will exercise constant supervision of the work while in progress, and will, to the best of his ability, construct and complete the work in a thorough and workmanlike manner throughout, it being the intention of this contract to secure the construction of said chimney in as perfect a manner as possible.

Contract Price.—1,250 dollars, or £260. 8s.

PHŒNIX BOLT AND NUT COMPANY, HANDSWORTH, BIRMINGHAM.

Architect, YEOVILLE THOMASON; *Builders,* W. PARKER & SON. *Built,* 1871.

	No. 1.	No. 2.
Description, brick	Octagonal.	Octagonal.
Total height, including foundation . . .	120' 0"	100' 0"
Height from ground line	110' 0"	90' 0"
Outside measurement at foundation . .	20' 0"	16' 0"
,, ,, at ground line . .	11' 0"	8' 2"
Inside ,, ,, ,, . .	5' 6"	3' 6"
Outside ,, at top	6' 6"	4' 6"
Inside ,, ,, ...	5' 0"	3' 0"
Foundation bed	Marl.	Marl.

SURREY COMMERCIAL DOCKS, ROTHERHITHE.

Engineer, JAS. A. McCONNOCHIE; *Builders,* HOLLAND & HANNEN. *Built,* 1883; April to August inclusive.

Description.—Square brick pedestal with panelled sides, circular brick shaft.

Dimensions,—

Height, including foundation	132' 2"
,, from ground line	108' 8"
Outside measurement, concrete bed, square	19' 6"
,, ,, brick footings ,, ...	17' 6"
,, ,, at ground line ,, ...	13' 9"
,, diameter at base of shaft, circular	10' 0"
Inside ,, ,, ,, ...	5' 6"
Outside ,, at top, under cap	7' 3"
Inside ,, ,, ...	5' 9"

Foundation.—The foundation bed is gravel 23' 6" below ground line. A square column of cement concrete 19' 6" × 19' 6" is carried to within 4' 6" of ground line, at which height the footings commence.

Fire-brick Lining.—A fire-brick lining is built up 34' 6" in height, in contact with brickwork of chimney proper but not bonded to same.

Construction.—The chimney has a square pedestal with panelled sides, brickwork 3' thick, 33' 4" high from bottom of brick footings to top of Portland stone moulding at base of shaft.

The shaft is built in Old English bond with all headers on external face of circular work, and is constructed in four sections, with thicknesses as follows :—

1st section	23' 11" high	1' 10½" thick.
2nd „	20' 0" „	1' 6" „
3rd „	20' 0" „	1' 1½" „
4th „	13' 4" „	9" „

The brickwork is then stepped out and surmounted by a moulded Portland stone cap 1' 10" deep. The stones forming cap are held together by a wrought iron strap.

Scaffold.—Outside.

Batter.—4½" in 20', or 1 in 53.

Capacity.—The fumes from No. 8 boilers are carried away by this shaft, and it is intended to serve engines of 430 h.p.

Lightning Conductor.—Copper tape.

Cost, including fixing, £17. 13s. 3d.

SOUTH METROPOLITAN GAS WORKS, RETORT-HOUSE SHAFT, OLD KENT ROAD, LONDON.

Figs. 65 to 68 inclusive.

Engineer, G. LIVESEY, M.I.C.E.; *Contractor,* — MORELAND, Old Street, St. Luke's.
Built . . . . . 1862.

Description.—Square brick shaft, built parallel the entire height and buttressed on the four sides, as see illustrations.

Dimensions,—

Height, including foundation	116'	0"
„ from ground line to top	108'	0"
Concrete foundation 17' 0" square ×	5'	0"
Outside measurement over buttresses at ground line . .	12'	0"
„ „ over square shaft „ „ . .	7'	4"
Inside „ of „ „ „ „ . .	5'	0"
Outside „ at top	6'	6"
Inside „ „	5'	0"

Foundation Bed.—Gravel.

Bond and Fire-brick Lining.—The shaft is built in English bond, with stock bricks outside and fire-bricks inside, bonded together; thus the courses of fire-bricks are alternately 9" and

4¼". The fire-bricks were specially obtained for this chimney, of the same size as the stock bricks. Mr. Livesey has sometimes adopted the plan of bonding the fire-bricks into the stocks every sixth to eighth course, where in using the ordinary sized fire-bricks the courses of the two coincide.

Bricks.—Total, about 80,000.

Iron Bond.—Bonds of 2" × ¼" flat iron, riveted at points of intersection, were built in the brickwork at intervals of 5′ in height. (See Fig. 68.)

Draught.—The pressure at the base of this shaft when only forty-four furnaces were in action and a quantity of cold air entered through the dampers of the remaining ten was equal to 0·95" of water, and the quantity of heated air and gases discharged was about 470 cubic feet per second.

Duty.—This chimney when erected had fifty-four furnaces working into the shaft, and it was intended to connect twenty more. The shaft has been in constant use since its erection in 1862, and the brickwork has not burned away. Mr. Livesey's opinion is that the heat is not sufficient to injure good fire-brick.

Cap.—Cast iron, weighing 2 tons.

Copper Rim.—On the top and sides of the 2′ 3″ blocking course, above cap, there is a copper rim ⅛" thick, put on at the suggestion of the contractor.

Lightning Conductor.—Copper wire.

Cost.—Complete, £530.

Remarks.—In a similar chimney 70′ high, built entirely of fire-brick for the engine-house, since the one above described, Mr. Livesey carried up the 14″ work about 50′ and then reduced it to 9″ work to the top by a set off inside; thus the flue is larger at top than at base.

TAMWORTH WATERWORKS.

Architect, ALEX. PAYNE, F.R.I.B.A., in conjunction with the *Engineer*, H. J. MARTEN, M.I.C.E.
Builder, JOHN GARLICK, Birmingham. *Built*, 1880.

Description.—Brick, circular pedestal and shaft.

Dimensions,—

```
Height above concrete foundation . . . . . . . . .  14'  6"
       „        „    ground line . . . . . . . . . . . 107'  0"
Outside diameter, circular pedestal . . . . . . . . .  11'  6"
Inside     „         „        „      . . . . . . . .    6'  6"
Outside    „      base of shaft . . . . . . . . .      8'  6"
       „     „      top    „     . . . . . . . . .      6'  3"
Inside    „     „       „      . . . . . . . . .        4'  0"
```

Foundation.—Massive concrete foundation.

Construction.—The chimney above concrete is built up in eight sections as follows:—

			Height.
1st section	Brick footings in mortar		4' 6"
2nd „	Brick pedestal „ 11' 6" external diam.		28' 0"
	Shaft.		
3rd „	Base mouldings in cement		5' 0"
4th „	Brick in cement, 8' 6" external diam. . . .		10' 0"
5th „	„ mortar, 8' 0" „ „ . . .		20' 0"
6th „	„ cement, 7' 6" „ „ . . .		10' 0"
7th „	„ mortar, 7' 0" „ „ . . .		17' 0"
8th „	„ cement, including cap		20' 0"
	Height above concrete . . . 114' 6"		

The shaft has one internal set off.

Hoop Iron Bond.—Hoop irons were used in several courses close together and several lengths in each course.

Brick in Cement.—The architect had the cement rings constructed "to give greater strength to the shaft and to obviate cracks, cement having greater adhesive power than ordinary mortar."

Fire-brick Lining.—Constructed 50' high.

Cap.—Constructed on a good ring of York stones well cramped together.

Lightning Conductor.—Copper tape, insulated.

BOSTON, LINCOLNSHIRE, FLOUR MILL CHIMNEY SHAFT.

Engineer, W. H. WHEELER, C.E., Borough Surveyor. *Built,* 1869-70.

Description.—Octagonal brick pedestal, circular shaft and cast iron cap.

Dimensions,—

Height from bottom of piles to top. . . . . 135' 3"
  „   „   „   concrete foundation to top. . . 119' 3"
  „   „   „   brick footings   „   . . . 111' 3"
  „   „   ground line to top . . . . 105' 3" .
Outside measurement over sides of pedestal at ground line 11' 6"
Inside   „   of fire-brick shaft . . . . . . 4' 9"
Outside   „   base of circular brick shaft . . . 8' 0"
Inside   „   „   „   „   „   . . . 3' 6"
Outside   „   at top, under cap . . . . . . 4' 6"
Inside   „   „   „   . . . . . . 3' 0"

Foundation Bed.—The upper part of ground is silty clay, and the lower part hard clay and chalk stones.

Construction.—The chimney is built on No. 9 piles, 15' long by 12" square. The shoes of piles are 30' below floor line of mill. The piles at top are connected together with capsills, on which rests cross planking, then a concrete bed 15' × 15' × 8', on the top of which the brick footings commence.

The octagonal pedestal is 20' high above floor level of mill and has Bramley Fall stone quoins.

The circular shaft has a stone base moulding 3' 9" in height. The first 3' of shaft at base above moulding and the first 9', measuring from under cap and mouldings at top, are built in cement.

The cap is of cast iron 1" thick, 9" wide, 9" deep inside and 3" deep outside, 4' 6" external diameter; constructed in three segments, bolted together with internal flanges.

Materials.—Brickwork, 16 rods. Bramley Fall stone, 740 cubic feet.

Fire-brick Lining.—The inner shaft of fire-brick is 40' high from ground line, built in 4½" work.

Cost.—£350.

A. GORDON & CO., CALEDONIAN ROAD, BREWERY, LONDON, N.

Fig. 77.

Engineer, C. PANTING; *Builders,* WILLIAMS & SONS.   *Built,* 1867; 3 to 4 months occupied in construction.

Description.—Octagonal brick pedestal, circular brick shaft.

Dimensions,—

Height, including foundation . . . . 114′ 0″
  ,,   from ground line to top . . . . 100′ 0″
Outside measurement, square concrete foundation 4′ 0″ thick 18′ 0″
   ,,   ,,   base of brick footings . . . . 14′ 0″
   ,,   ,,   at ground surface . . . . 11′ 3″
Inside   ,,   of fire-brick lining . . . . 3′ 6″
Outside   ,,   at base of shaft . . . . 7′ 3″
Inside   ,,   ,, ... 3′ 6″
Outside   ,,   at top, over C.I. cap . . . . 4′ 7″
Inside   ,,   ,, ... 3′ 0″

Footings.—The footings are stepped upwards from 14′ square to 11′ 7½″, in seven courses; from this height the footings in thirteen courses are stepped at the four corners until they work into the octagon pedestal at ground line.

Pedestal.—Octagonal, panelled on each face, and 25′ high from ground line, including a stone cornice, from which starts the circular shaft.

Shaft.—The shaft is constructed in four sections with thicknesses as follows :—

1st section . . . . 21′ high 1′ 10¼″ thick.
2nd ,, ... 15′ ,, 1′ 6″ ,,
3rd ,, ... 15′ ,, 1′ 1½″ ,,
4th ,, ... 24′ ,, 9″ ,,

Batter.—Shaft 10′ vertical, 15′ 1 in 60, 50′ 1 in 42 nearly.

Cap (Fig. 77).—The shaft has a cast iron terminal 5′ 6″ high, of ¼″ metal, made in segments and bolted together by internal flanges. It was painted in three coats of good oil colour, the last coat being Portland cement in oil.

Cost.—£440.

DONCASTER LOCAL BOARD OF HEALTH, SEWAGE IRRIGATION
WORKS.

Engineer . . . . B. S. BRUNDELL.

Description.—Square brick pedestal, surmounted with square panelled shaft.

Dimensions,—

Total height, including foundation				112'	9"
Height from ground to top				100'	0"
Outside dimension at foundation, square				18'	0"
,,	,,	at bottom of footings, square		14'	3"
,,	,,	at ground line		9'	9"
Inside	,,	,,		3'	11"
Outside	,,	at commencement of shaft		7'	9"
Inside	,,	,,	,,	3'	11"
Outside	,,	under cap		6'	1"
Inside	,,	,,		3'	9"

Pedestal.—This is 17' high, 2' 5" thick brickwork, with stone moulded plinth at base and moulded stone cornice at top

Shaft.—Square with panelled sides, erected in three sections as follows:—

1st section . . . 25' high	18" brickwork.
2nd ,, . . . 25' ,, ...	14" ,,
3rd ,, . . . 22' ,, to under cap	9" ,,

Batter.—1 in 86·4.

Cap.—Ornamental stone top, surmounted by a cast iron cap.

GREAT NORTHERN RAILWAY. FARRINGDON STREET GOODS STATION.

Engineer, R. JOHNSON, M.I.C.E.; *Builders*, KIRK & RANDALL.
Built, 1880; five weeks occupied in construction.

Description.—Round brick shaft on square pedestal.

Dimensions,—

Total height, including foundations			102'	6"
Height from ground line to top			96'	6"
,, of pedestal, including footings and stone base moulding to shaft.			23'	9"
Outside measurement at ground line, square			8'	6"
Inside	,,	,, circular	3'	0"
Outside	,,	base of shaft	7'	6"
Inside	,,	,,	3'	0"
Outside	,,	under cap	4'	6"
Inside	,,	at outlet top of cap	2'	9"

Foundation.—Concrete base 12' 6" square × 3' deep on foundation bed of London clay.

Materials and Weight,—

	Tons.	Cwt.
In base there are 22,000 bricks, weight, including mortar.	86	0
,, shaft ,, 21,000 ,, (radiating) ,, ,,	64	0
,, base, weight of stone	11	0
,, cap ,, ,, ...	1	16
,, ,, cast iron terminal of ⅜" metal, in segments, bolted together by internal flanges	0	12
	163	8

Beart's patent perforated radiating bricks were used in the circular shaft, laid all headers on external face.

Scaffold.—Inside and outside.

Construction.—The walls of pedestal are 2' 9" thick, shaft is built up in five sections, as follows :—

1st section	9' high 2' 3" thick.
2nd ,, ...	16' ,, 1' 10½" ,,
3rd ,, ...	16' ,, 1' 6" ,,
4th ,, ...	16' ,, 1' 1½" ,,
5th ,, ...	top . 9" ,,

Fire-brick Lining. — This is carried up independent of pedestal, with about 1" space for expansion, to a height of 21' 6" in 4½" work set in fire clay.

Batter.—The first 9' of circular shaft is built vertical, then 16' to a batter of 1 in 42, and the topmost length of 41' to underneath cap 1 in 37 nearly.

Duty.—To carry off the products of combustion from flues of No. 2 Cornish boilers 24' long × 6' external diameter, each having one internal tube 3' diameter.

Lightning Conductor.—Copper tape.

Cost.—Chimney complete, £400.

MESSRS. PEEK, FREAN & CO.'S CHIMNEY, DRUMMOND ROAD, LONDON.

Architects, SNOOKE & STOCK ; *Builders,* RIDER & SON.
Built, 1866 ; two months occupied.

Description.—Square brick shaft and pedestal.

Dimensions,—

Total height, including foundation	99'	3"
Height from ground line	91'	3"
Outside measurement at ground line	10'	0"
Inside ,, ,, ,, ...	4'	9"
Outside ,, at top	7'	3"
Inside ,, ,, ...	4'	6'

Foundation Bed.—Sandy gravel.

Batter.—1 in 66.

Bricks.—66 *m.* used in construction.

Weight.—313 tons.

Scaffold.—Outside, costing £60.

Lightning Conductor.—Stranded copper rope.

Cost.—Chimney complete, £560.

EXPERIMENTAL CHIMNEY, MAYFIELD PRINT WORKS, MANCHESTER.

Engineer, R. ARMSTRONG; *Owners,* T. HOYLE & SONS.
Builders . . . D. BELLHOUSE & SONS.

Description.—Brick, square foundations and pedestal; shaft octagonal.

Dimensions,—

Height, including foundations	. . . .	100'	0"
,, above ground line	. . . .	90'	0"
,, ,, level of fire-bars	. . . .	88'	0"
,, of pedestal above ground	. . . .	18'	0"
,, of shaft, including cap	. . . .	72'	0"
Outside measurement base of footings	. . . .	15'	0"
,, ,, square pedestal	. . . .	8'	0"
Inside ,, ,, ,, octagonal flue	. .	5'	0"
Outside ,, base of octagonal shaft	. . . .	8'	0"
Inside ,, ,, ,, ,,	. . . .	4'	10"
Outside ,, top of ,, ,,	. . . .	4'	4½"
Inside ,, ,, ,, ,,	. . . .	2'	10½"

Shaft.—This is octagonal both inside and outside, and is built up in two sections, viz. :—

1st section . . . . 36' high 19" thick.
2nd ,, ... 36' ,, 9" ,,

Batter.—1 in 36.

Capacity, &c.—The internal capacity of the chimney, together with the short flue connecting it with the boiler to which it is attached, is about 50 cubic yards.

The internal horizontal area of the outlet—the narrowest part—is about 1,000 square inches, which, as it was intended to make the shaft large enough for about 50 h.p., gives 20 square inches for each h.p., while there is about 1 cubic yard capacity in the chimney for each h.p.

These proportions were adopted after averaging some of the best chimneys the engineer and proprietors were acquainted with in Lancashire.

*Cost.—£*100.

GREAT NORTHERN RAILWAY WOOL SHED, BRADFORD.

Engineer, R. JOHNSON; *Builder,* W. PICKARD & SON. *Built,* 1875.

Description.—Square stone shaft and pedestal.

Dimensions,—

Height, including foundation	. . . .	96'	0"
„	above ground line	90'	0"
„	of pedestal above ground line	19'	0"
Outside measurement at foundation	. . . .	12'	0"
„ „	at ground line	9'	0"
Inside „	„ „	2'	9"
Outside „	above pedestal	6'	9"
Inside „	„ „	2'	9"
Outside „	under cap	4'	10"
Inside „	at top	2'	6"

Fire-brick Lining.—A 4½" fire-brick lining is constructed independently of the stone work to chimney, for the height of the pedestal.

Duty.—No. 4 boilers, each 24' long; outside diameter, 6'; internal flue diameter, 3'.

Batter.—Pedestal, parallel. Shaft, 1 in 66.

*Cost.—£*260.

Specification, Extract from.—The site is on rock, and if in digging out for the foundations it is found that any faults or defective places exist, or if the rock has been quarried to a greater depth than that shewn for the footings, then the excavation is to be continued down to the solid, and a wall 1' wider than the lowest course of footings is to be built up to the proper level for commencing the walls. This wall is to be of rubble masonry in Portland cement, or otherwise (as the Engineer may determine) of concrete composed of stone, broken to pass a 2" mesh, sand and Portland cement in the proportion of six of stone, two of sand and one of cement, to be properly mixed together with the least quantity of water, and thrown into the trench from a height of at least 8'.

The mortar is to be composed of the best Skipton or South Elmsall lime, as the Engineer may determine, and clean sharp sand, in the proportion of one of lime to two of sand, to be ground together in a mill worked by steam power on the ground, and only prepared as required for use.

The walls are to be built with block in course, pitched face stone from the Bradford, Idle or Calverley quarries, as the Engineer may determine; the stones to be not less than 11″ on the beds and 6″ on the joints, and not less than 7″ or more than 14″ in thickness; the thicker courses at the bottom, and diminishing gradually upwards; thick courses in no case to be laid upon thin ones; the whole to have boasted beds and joints. No natural faces must show in the wall. The footings are to be of rag stone.

The chimney is to be carried up true and vertical upon its axis; the base is to be lined inside with Stourbridge fire-brick to a height of 20' from bottom of flue. The shaft is to have an internal width at base of 2' 9″ square, and at top 2' 6″ square. At junction of shaft with pedestal it is to be corbelled over internally 6″ all round. It is to be 2' thick at this point, and is to be gradually diminished to a thickness of 1' 3″ under cap. A space of 4″ is to be left between the top of fire-brick lining and the bottom of corbelling in chimney, to allow of free expansion.

The base moulds, necking and cap to chimney, together with all oversailings, strings and dentils, are to be executed in strict accordance with the drawing upon which all the sizes and forms are given. The cap and base mouldings are to be set in cement, the profiles of which are to be cleanly and sharply cut.

The fire-bricks are to be of the best Stourbridge, or quality equal thereto, set in ground fire-clay, mixed with water to the consistency of cream. The bricks are to be dipped into the liquid fire-clay, laid in place and hammered together, so as to be when finished brick and brick.

KENT WATER WORKS.

Fig. 80.

Engineer, W. MORRIS, M.I.C.E.; *Builders,* J. RIDER & SON.
Built, 1879; four months occupied in construction.

Description.—Brick, square pedestal, circular shaft, stone cap.

Dimensions,—

Height, including foundation	97'	3"
,, from ground line to top	90'	0"
Outside measurement at ground line	9'	3"
Inside ,, ,, ,, ...	4'	0"
Outside ,, at base of circular shaft	7'	9"
Inside ,, ,, ,, ,, ...	4'	0"
Outside ,, at top, under cap	5'	6"
Inside ,, ,, ,, ...	3'	3"

Foundation Bed.—Loamy gravel on chalk.

Pedestal.—22' high with Portland stone mouldings, 1' 10¼" thick brickwork.

Fire-brick Lining.—20' high from ground line, 4½" thick; no space between brickwork of pedestal and lining.

Shaft.—This is erected in three sections, viz.:—

1st section	. . . 21' high	. . . 1 10½"	thick.
2nd ,,	. . . 25' ,,	. . . 1 6"	,,
3rd ,,	. . . 22' ,,	with cap 1 1½"	,,

Wrought Iron Bond, &c.—No. 11 rings of wrought iron bond are built into shaft at about 6' intervals, and No. 1 wrought angle iron band placed round circumference of Portland stone cap; also No. 2 copper rings built into stone cap.

Batter.—1 in 56.

Scaffold.—Outside.

Lightning Conductor.—Copper tape.

Duty.—No. 6, 35 h.p. boilers connected to shaft.

G. TUCKER & SON, BRICKWORKS, &C., LOUGHBOROUGH.

Designed and built by G. TUCKER & SON.

Description.—Square brick.

Dimensions,—

Height, including foundation	95'	0"
,, from ground line to top	85'	0"
Outside measurement at foundation	12'	0"
,, ,, at ground line	8'	0"
Inside ,, ,, ...	4'	0"
Outside ,, at top	5'	6"
Inside ,, ,, ...	4'	0"

Foundation.—Concrete on marl foundation bed.

Bricks.—Ordinary size, 9″ × 4½″ × 3″.

Batter.—1 in 68.

Scaffold.—Inside.

Cap.—Constructed of large purposely-made bricks.

Lightning Conductor.—Galvanised iron strand.

Duty.—To carry off smoke, &c., from two boilers and a Hoffman kiln.

Cost.—About £120.

WEST END LAUNDRY COMPANY'S CHIMNEY, FULHAM, LONDON.

Architect, WILLIAM C. STREET; *Builders,* WEST END LAUNDRY COMPANY.
Built, 1883 (spring of year); time occupied, four months.

Description.—Brick, circular shaft, on square pedestal.

Dimensions,—

Total height, including foundations		. . . .	97′	6″
Height from ground line		. . . .	80′	0″
Concrete foundation, square		. . . .	17′	6″
,, ,, in height		. . . .	6′	0″
Outside measurement base of footings		. . . . . .	13′	3″
,, ,, bottom of square pedestal		. . .	10′	3″
,, ,, top ,, ,,		. . . .	9′	9″
,, ,, base of circular shaft		. . . . .	8′	0″
Inside ,, ,, ,, ,,		. . . . .	4′	3″
Outside ,, top of shaft		. . . .	5′	0″
Inside ,, ,, ,,		. . . .	3′	6″

Foundation.—The foundation bed is gravel, 17′ 6″ below ground level.

Fire-brick Lining.—The 9″ fire-brick lining extends 22′ from inside base of pedestal, and is built up square 2′ 6″ × 2′ 6″, leaving a cavity at base of 6″ and top 4″ between fire-brick and inside of pedestal.

Pedestal.—This is 27′ 6″ high from top of concrete, and has a batter of 1 in 52.

Construction,—

Pedestal	27′	6″ high 2′	7½″	brickwork.	
Shaft 1st section .	16′	0″ ,,	1′ 10½″	,,	
,, 2nd ,, .	16′	0″ ,,	1′ 6″	,,	
,, 3rd ,, .	16′	0″ ,,	1′ 1½″	,,	
,, 4th ,, .	16′	0″ ,,	9″	,,	

91′ 6″ height from base of footings.

Bricks.—59 *m.* ordinary bricks used in construction, laid to Flemish bond.

Scaffold.—Outside.

Weight.—301¼ tons, giving a little less than one ton pressure per super foot on foundation bed.

Duty.—One boiler flue and one vapour flue connected to shaft; total h.-p. 45.

HASTINGS ELECTRIC LIGHT COMPANY'S CHIMNEY.

Architect, E. W. J. HENNAH; *Builder,* A. VIDLER.
Built, 1882-3 (August to March).

Description.—Square brick.

Dimensions,—

Total height, including foundation . . . . 85′ 0″
Height from ground line . . . . 78′ 0″
Outside measurement at ground line . . . . 6′ 6″
Inside   ,,   ,,   ,, ... 3′ 6″
Outside   ,,   at top . . . . 4′ 0″
Inside   ,,   ,, ... 2′ 6″

Foundation.—Concrete on clay and sandstone.

Batter.—The sides for the first 27′ of shaft are parallel; for the remainder of height they diminish to a batter of 1 in 41.

Materials.—30 *m.* of ordinary bricks used in construction laid in cement. Measured work, 6 rods, 226′ reduced brickwork. One course 4″ stone on top.

Weight.—About 90 tons.

Scaffold.—Outside, costing about £17 extra.

Duty.—Two boilers working up to 200 h.p.

Lightning Conductor.—Copper tape, cost £12 fixed.

Cost.—Chimney and scaffold complete, £147.

MESSRS. ROBERT HEATH & SONS' CHIMNEY SHAFT, RAVENS-
DALE IRON WORKS, TUNSTALL, STAFFORDSHIRE.

Description.—Circular wrought iron shaft, not spread at base.

Dimensions,—

Height from ground line to top	75' 0"
Outside diameter at ground line	6' 0"
,, ,, at top	6' 0"

Wrought Iron Plates.—No. 75 wrought iron plates were
used in the construction of this shaft, the thickness being ¼".
The plates have a lap of 2¼", and are riveted together with ⅞"
cup-headed rivets.

Fire-brick.—The shaft is lined its entire height with fire-
brick.

Duty.—The shaft carries off the fumes from three boilers.

GEO. M. HAMMER & COMPANY, CROWN WORKS, BERMONDSEY,
LONDON.

Architect and Builder, G. M. HAMMER & CO.; *Built,* 1882.

Description.—Brick, square pedestal, octagonal shaft.

Dimensions,—

Height, including foundation	80' 0"
,, from ground line to top	70' 0"
Portland cement, square concrete foundation 5' thick . .	14' 0"
Height of pedestal	15' 0"
Outside measurement at ground line	7' 4"
Inside ,, ,, ...	2' 9"
Outside ,, at base of octagonal shaft . . .	5' 10"
Inside ,, ,, ,, . . .	2' 10"
Outside ,, at top, under cap	4' 0"
Inside ,, ,, ,, ...	2' 6"

Foundation Bed.—Ballast.

Shaft.—This is built up in three sections, as follows :—

1st section	15' high 18" thick.
2nd ,, ...	20' ,, 14" ,,
3rd ,, ...	20' ,, 9" ,,

Scaffold.—Outside.

Batter.—Pedestal, parallel. Shaft, 1" in 5' or 1 in 60.

I

Weight.—90 tons including footings = ·45 tons per square foot on concrete foundation.

Lightning Conductor.—Copper strand rope.

Duty.—No. 1, 30 h.p. boiler now connected, and the proprietors consider it capable of taking another 30 h.p. boiler.

Cost.—£208 complete.

MESSRS. HARVEY & SON, BREWERY CHIMNEY, LEWES.

Fig. 69.

Architect, W. BRADFORD ; *Builders,* H. CARD & SON.   *Built,* 1881 ; about two months occupied in construction.

Description.—Octagon brick on square pedestal.

Dimensions,—

Total height, including foundations	. . . .	85′	0″
Height from ground line	. . . .	70′	0″
Outside measurement at ground line, square	. . . . .	6′	6″
Inside ,, ,, ,, ,,	. . . . .	2′	0″
Outside ,, at top	3′	6″
Inside ,, ,, square	2′	0″

Foundation.—Blue lias concrete, 17′ × 16′ × 8′ 6″ deep, on clay foundation bed.

Scaffold.—Outside.

Cap, &c.—Stone in cap and base, cost £70.

Duty.—To carry away smoke, &c., from one boiler of 25 h.p.

Lightning Conductor.—Copper rope, cost £10 fixed.

Cost.—Chimney complete, £270.

CHAIN, CABLE AND ANCHOR TESTING WORKS, RIVER WEAR COMMISSIONERS, SUNDERLAND.

Figs. 70 *and* 71.

Engineer, H. H. WAKE, M.I.C.E.; *Builder,* GEO. GRAINGER.
Built, 1873, at the rate of 1′ 6″ in height per day.

Description.—Concrete, square pedestal, octagonal shaft.

Dimensions,—

Height, including foundation	62'	6"
„ above ground line	56'	6"
„ of pedestal above ground line	24'	9"
„ of shaft, including base moulding	31'	9"
Concrete foundation square	12'	0"
Outside measurement at ground line	7'	6"
Inside „ „ „	4'	0"
Outside „ base of shaft	5'	9"
Inside „ „ „	3'	3"
Outside „ at top	3'	4"
Inside „ „	1'	10"

Foundation.—The foundation bed is "forced ground," consisting of sand and town rubbish tipped on to sea beach; upon this is laid a concrete bed $12' \times 12' \times 6'$.

Pedestal.—The square pedestal is outside $7' 6" \times 7' 6"$ parallel. Inside to a height of 19' the flue is 4' 0" square and parallel; from this height (see Fig. 71) the insides of pedestal have a batter similar to that of the insides of octagonal shaft.

Shaft,—

Base moulding .	1' 9"
Shaft and cap .	30' 0"

The shaft is built up in two sections, as follows:—

1st section	15' high 1'	3" thick.
2nd „	15' „	9" „

Fire-brick Lining.—The lining extends the whole height of pedestal, viz., 24' 6" above ground, and is constructed in two thicknesses:—

15' high 9" thick.
$9\frac{1}{2}'$ „ $4\frac{1}{2}"$ „

There is no cavity or space between the fire-brick and the concrete of pedestal.

Batter.—1 in 25.

Materials.—Concrete to foundation and pedestal was composed of 1 of Portland cement to 8 of shingle and sand. Concrete to shaft, 1 Portland cement to 5 gravel and sand. Outside of chimney was coated with $\frac{1}{4}"$ of cement, in the proportion of 1 cement to 1 sand, and joints struck to represent Ashlar.

Construction.—The shaft from base moulding was constructed in the following manner:—Wood moulds 3' in height, formed of $\frac{3}{4}"$ boards, hinged together in pairs at their outer edges, were constructed to form the quoins of the octagonal shaft. Standards 6' in height were used to bolt the outside moulds to the inside ones, and also to have 3' hold on the work already executed while a 3′ length was being constructed. The batter of the shaft being uniform the quoin moulds did not require re-adjusting every 3′ to suit the decreasing girth of the shaft, but this was provided for by wedge pieces being placed on each face working between the quoin moulds; these wedges had to be reduced in width every 3′. When the quoin moulds had reached half the height of the shaft their edges on each face met, and wedges could not be inserted. The moulds were, therefore, re-adjusted, and a fresh set of wedges started; the edges of quoin moulds met again at top.

TAKING DOWN SHAFT.

MESSRS. GILKES, WILSON, PEASE & CO., TEES IRON WORKS,
MIDDLESBOROUGH.

Figs. 72 and 73.

Engineer . . . . Mr. CHAS. WOOD.

TAKING DOWN A CHIMNEY SHAFT.

An ingenious arrangement for facilitating the taking down of an old chimney shaft was here employed.

In consequence of the shaft standing in a crowded position, the plan of letting it fall was inadmissible, and it had to be taken down from the top.

The bricks had to be lowered with as little damage as possible, so that they might be used again for building purposes. Owing to the position of the chimney the bricks could not be thrown down outside, and if thrown down inside they would have been smashed, and if lowered by mechanical means the process would have been very tedious.

Under these circumstances it was considered whether the bricks could not be allowed to fall by their own weight, and at the same time be cushioned sufficiently to break their fall and prevent damage.

In order to do this an air-tight iron box was placed at the bottom of the chimney (Fig. 72). This box was fitted with an air-tight door, mounted on hinges and closing on an india-rubber face, against which it was tightened by a wedge.

A wooden spout was then fixed on to the top of the box and carried up the chimney. It was 3½″ × 5″ inside, made of planks 1⅛" thick, well nailed together with a little white lead on the edges, thus making it air-tight. The spout was made in about 12' lengths, and these were joined together by cast iron sockets or shoes (see Fig. 73), and caulked round with tarred yarn, the whole apparatus costing about £6.

A few stays were put inside the chimney to keep the spout steady. Steps were nailed upon the wooden spout by which the workmen ascended. The whole of the spouting being air-tight, if a brick filled the spout it would not descend, but as the size of a brick is 3" × 4½" and the spout was 3½" × 5", there was ¼" space each side, through which the air could pass the brick freely, this space further allowing for any irregularity in the sizes of the bricks. The result was, that the bricks being cushioned in their fall, arrived at the base without damage.

As soon as the box was full the man at the bottom rapped on the spout as a signal to stop, and then opened the air-tight door and removed the bricks inside. This being done he shut the door and signalled "all right" to the man at top.

The workman at top lowered his own scaffold, and as the spout became too high he cut a piece off. If there were much mortar adhering to the bricks it was knocked off before putting the latter into the spout. Such mortar, &c., was allowed to fall inside the chimney and was afterwards wheeled out.

The plan here described was, we believe, quite novel, and is certainly simple. There are, no doubt, many similar circumstances under which it might be advantageously employed.

STRAIGHTENING SHAFTS.

CHIMNEY SHAFT, BINGLEY, NEAR BRADFORD.

Straightening.—This chimney, which was built at a cost of £2,000, was found some few years since to be 4' 6" out of the perpendicular. The inclination was found to be due to the foundations having settled under the superincumbent mass of brickwork at one side. In this emergency it was advised that excavations should be made under the foundations at the other side until the chimney settled down to the same extent, and so brought itself back into the perpendicular. A well was actually dug beneath the higher side of the shaft, and supplied with water to favor the yielding of the rigid part of the foundation. This was a very perilous method of straightening the shaft, as after the ground had become sodden with the water and the foundation of the shaft undermined it would have been next to impossible to have stopped the subsidence at the desired point, and the shaft would have most likely fallen in an opposite direction to that in which it had previously leant. Before, however, any injurious effect had occurred the rectification was placed in the hands of Messrs. Sanderson, of Huddersfield. The first step was the restoration of the foundation to its original state; the well was filled in and made firm with brickwork and concrete. A gap was next cut half through the bottom of the shaft on the side where no settlement had taken place by removing three courses of brickwork. As this was being effected strong screw-jacks of iron were inserted perpendicularly into the gap to take the pressure of the unsupported mass of brickwork. The jacks were 10" long, and iron plates 1" thick were inserted above and below their ends to serve as temporary platforms. The jacks were inserted one after the other, a few inches apart, as the brickwork was cut away, and each one was adjusted by its regulating screw, so as to come at once into right bearing for · sustaining its share of the weight. When the entire gap had been formed and all the screw-jacks inserted, the jacks were very slowly and gradually shortened as the weight did its work, and when the shaft had nearly settled back into its original perpendicular position, the portions of the gap, which lay between the screw-props, were filled in with masonry, the screws were, one after the other, removed, and masonry also put in their place. The making good of the work was commenced before the shaft had quite reached the perpendicular, because it was known that a slight allowance must be made for a small compression of the new work after the entire filling in of the gap. The straightening, as here detailed, was successful, and the shaft now stands perpendicular.

CHEMICAL WORKS, PITCHCOMBE, GLOUCESTERSHIRE.

Straightening Chimney. — The octagonal brick shaft at Messrs. Matthews & Sons' Works, as above, built 1862, and 132' high, gradually settled from its upright position until in 1875 it was found to be 3' 10" at top from the perpendicular. Mr. H. J. Taylor, of Nailsworth, assisted by three workmen, undertook to straighten the shaft, by means of cutting out one course of bricks on the five sides opposite to the inclination, inserting a thinner course in its place, and letting the stack regain its upright position by the action of gravitation. A platform was erected about 40′ from the base, and the walls, which at this height were 2′ in thickness, cut through by means of hammers and chisels. As the bricks were removed a thinner course was substituted, and the space above filled with iron wedges. This work lasted three weeks, the weather being most unfavourable. The chimney stood the operation well. When everything was in readiness the wedges were withdrawn, and the stack settled to within an inch or two of the perpendicular.

It had been calculated that $\frac{1}{4}''$ would bring the stack back 7″ at top, so that the difference in the thicknesses of the courses had to be $1\frac{5}{8}''$.

The cost of a new shaft was estimated at £800, and the old one was straightened at about one-tenth this amount, and is still working satisfactorily.

FALL OF SHAFTS.

STRAIGHTENING AND FALL OF A CHIMNEY.

Oldham.—A large shaft 165′ high, 16′ diameter at base and 7′ 6″ diameter at top, was, in 1873, constructed at the new works of Messrs. Abraham Stott & Son, Osborn Mills, Feather Stall Road, Oldham. It was found necessary to straighten the pile, which leaned considerably over to one side. The owners, therefore, entered into a contract with two brothers named Gradwell, of Newton Heath, to "saw" the chimney. Mr. Stott found that the men were taking out a whole course of bricks, at a third of the height of the shaft, and temporarily supplying its place with a series of wedges of wood and iron, instead of "sawing" the mortar out as arranged. The owner at once remonstrated with one of the men, but he declared the plan was safe, and took Mr. Stott to some rising ground adjacent, to observe the safety, when, as if in ridicule of his assertion, the chimney toppled over, except about 30′ at the base. The poor fellow on the scaffold was buried in the fallen bricks. Adjacent to the shaft was the boiler house, a large new building, which was completely wrecked, a portion of the stalk falling upon it.

At the inquest it was deposed that the shaft was built on sound principles, but in a faulty manner. The material was inferior, the bricks being soft and the mortar improperly mixed, consisting too largely of sand.

TAKING DOWN AND FALL OF A CHIMNEY.

Glasgow.—A chimney nearly 100' high in the yard of a railway wagon builder at Glasgow, fell on January 10th, 1870, killing two men. The stalk had been deemed insecure, and the men killed were employed with two others in taking down 20' from the top. Cross stays were erected inside the shaft, by means of which the men ascended to the top, where they removed the bricks, and dropped them inside the chimney. An aperture was made in the base 5' high by 3' wide, through which the two men at foot of shaft removed the fallen bricks. While they were so engaged the stalk suddenly gave way at the base, killing the two men at top; the men at the bottom, however, escaped without injury. The accident was supposed to have been occasioned by the aperture weakening the base.

HEIGHTENING AND FALL OF A CHIMNEY.

Bury.—On January 23rd, 1884, a chimney at Messrs. Allen and Parker's Eaton Vale bleachworks, near Bury, fell, killing three women and wrecking the building. The shaft was 105' high, 7' 2" at base, 4' inside and 9" brickwork at top. The chimney was originally 90' high and was partly pulled down, rebuilt, raised and the batter lessened in June, 1870, by Messrs. Chris. Hardman & Sons, to increase the draught. In December, 1882, Mr. Robert Williams, of Elton, straightened the shaft by taking out a course of bricks from one side and replacing it with a thinner one; he also put five wrought iron ties round the shaft, the uppermost one being above the cornice and stonework, there being a fissure in the blocking at top, part of which was nearly hanging off. The work was done in frosty weather. The crack was filled up with mastic and oil.

Mr. James Maxwell, an architect, who examined on behalf of the Coroner, stated at the inquest that the base was defective, not only in its area but in the strength of the walls and the construction of same. The walls were built of 9" brickwork and 9" stonework outside. Little attempt had been made to obtain a proper bond. The increased height, he considered, to be the primary cause of the accident.

The chimney was observed, at the time of accident, to be caught by a gust of wind that caused it to vibrate about a foot, and a second gust following brought down the stalk. It appeared to break about 30′ from the top. It fell in the direction to which it had previously inclined.

A new shaft was to have been commenced on the day of accident.

VIOLENT GALE, 1873, AND FALL OF CHIMNEYS.

Sheffield.—A square chimney, 116′ high above ground, in Trippet Lane, Sheffield, belonging to Mr. W. Reynolds, fell during a severe gale on December 16th, 1873, killing 10 persons and injuring five. The chimney in its fall demolished a four-storey building adjoining. The chimney had often been observed to sway when the wind was high, and on the morning of the accident a number of people were watching the oscillations, but the workpeople took no notice of it. The top was ornamental and made of terra cotta; the cornice was hollow, so as to be as light as possible; there was also a balustrade on the top. Mr. T. H. Jenkinson, an architect, gave evidence at the inquest that after the chimney had been erected exaggerated reports were circulated as to its rocking, and he made an examination during a severe gale and found the oscillation amounted to 2″ each way. The attention of the Town Council had, he believed, been called to the chimney a few years ago. The chimney tapered about 1′ from the base to the top. Mr. W. Reynolds, builder, who also owned the shaft, said he was responsible for the erection; nothing could have been put together better. It was built on the solid rock 4′ below ground, and was 116′ high from ground line. A quarter of an hour before the chimney fell he observed it oscillating more than it usually had done. The chimney broke 50′ above the base. When the complaints were made to him soon after the completion he greatly strengthened the base, and thereby prevented much oscillation. The outside was composed of the best pressed bricks.

The shaft was constructed in 1858 and cost £300.

Eight workshops were destroyed, and considerable damage done to the 20 h.p. engine by the fall.

The damage to property, irrespective of a large stock of cutlery, in process of manufacture, was estimated at £3,500.

At the inquest the jurors expressed their opinion that the shaft was too high for the width of the base.

The gale causing this disaster was an exceptionally severe one, and also caused the demolition of the following shafts in the neighbourhood, viz. :—

Portobello.—Messrs. Chris. Johnson & Co., cutlery manufacturers, Western Works, two chimney stacks blown down. No injury except to buildings.

Wicker.—Mr. Freckingham, Willey Street, engine chimney to the mortar mills blown down, falling through mortar mill roof.

Spital Hill.—Mr. J. Blyde, Hallcar Works, chimney stack fell, completely smashing roof.

Bolsover Street.—Mr. J. Dodworth, shaft to engine blown down.

Furnace Hill.—Messrs. Longden & Co., Phœnix Foundry, shaft 60′ high blown down.

Watery Lane.—Messrs. W. Guest & Sons, Neptune Works, shaft blown down, demolishing a new cutlery shop and damaging house adjoining.

Milton Street.—Messrs. Matthewman & Sons, Milton Works, large brick chimney fell, considerably damaging the works.

Saville Street.—Messrs. Thos. Firth & Sons, Norfolk Works, shaft blown down, injuring three workmen and causing great destruction of machinery and buildings. The chimney was 120′ high with 1′ 6″ walls, and snapped about the middle. The falling bricks broke the steam pipes from the boilers, and the steam severely scalded one of the men; he was also crushed about the legs and body. Damage to buildings and machinery was estimated at £1,000.

BLOWING-DOWN SHAFTS.

BLOWING-DOWN CHIMNEY.

Dublin.—On April 10th, 1884, an attempt was made by the Royal Engineers to blow down a conical shaft built 1820, at the works of the Glass Bottle Company, North Lotts, Dublin, in the occupation of Mr. W. Campbell. In compliance with a requisition sent to the military authorities trained men were brought from the Curragh, and a dynamo-battery sent from Chatham to fire the charges. The work was under the superintendence of an officer of the Royal Engineers. The shaft was a truncated cone 95′ in height and 64′ diameter at base, with walls 4½′ thick, tapering to 1½′ at top, and the weight was roughly estimated at

2,500 tons. The shaft rested upon twelve piers or pillars of brickwork; in each of these a charge of cannon powder weighing 2¾-lbs. was placed, closely "tamped up" and connected with a double electric wire attached to a low-pressure dynamo-machine, so as to fire the charges simultaneously. The machine was some 150 yards distant, on rising ground. At noon the charges were fired and a dozen little puffs of smoke became visible, there being little or no concussion, and with the exception that on the north and south sides of the shaft portions of brickwork slid slowly to the ground, the chimney stood as erect as before. A second attempt was made with 20-lbs. of gunpowder, which was placed under one of the smaller arched openings on a beam that traversed it, and was closely packed with sand bags. The wires were connected afresh and the charge was exploded, but with no result beyond destroying the sand bags. At the suggestion of a bystander a chain was attached round one of the brick piers and the end fastened to the axle of a dray drawn by a couple of horses, but these failed to secure the desired effect and the idea was abandoned. An engine of the Great Southern and Western Railway, a branch of which adjoined the works, was next called into requisition. A chain was attached from the engine round a beam that spanned an opening in the base, the brickwork near to which looked shaky, the only result being that the beam broke. The chain was next passed round one of the brick supports and this time the draw bar of the engine gave way. The chain, however, was next attached to the rear of the engine and this time the chain broke. A 3″ diameter hawser was then attached and the engine started, but the hawser, imperfectly fastened, dropped off. The men were again approaching the stubborn mass when there occurred a rumbling sound and a cloud of dust, and the immense shaft collapsed, fortunately before the men had reached the chimney.

Warrington, Lancashire.—About nine years ago the tall circular brick shaft at Messrs. Muspratt's Chemical Works, Warrington, 406′ high, 46′ diameter at base and 17′ diameter at top, was blown down by gunpowder, the works having been moved to another locality and the chimney therefore being in disuse. Mr. Stephen Court, engineer and architect to the St. Helen's Canal and Railway Company, superintended the demolition. A number of holes were delved round the base and fourteen charges of gunpowder inserted. The train was fired at 2.30 p.m. Nine charges exploded without any apparent damage being done to the stability of the shaft, but the report of the tenth had no sooner been heard than the chimney was rent from top to bottom and the huge mass gradually disintegrated from the base upwards. The whole of the stalk fell nearly within the circumference of its own base. No accident occurred.

MOVING SHAFTS.

MOVING A CHIMNEY.

Cabot Company's Cotton Mill, Brunswick, Maine, U.S.A. A shaft at the above works was moved 20′ in May, 1872, to allow of the enlargement of the mill. The shaft was 78′ high, 7′ 9″ square at base, 5′ square at top, contained 40,000 bricks, and weighed about 100 tons. The work was accomplished by moving the chimney on planed and greased planks by means of two screw jacks. The flues were re-connected, and the fires started within 8½ hours from the commencement. The removal was planned and executed by Mr. Benjamin Greenes.

CLIMBING CHIMNEYS.

TALL CHIMNEY CLIMBING.

Messrs. Sanderson & Co., of Huddersfield, have an ingenious method of ascending to the tops of tall chimneys for purposes of examination and repair. It consists in pushing length after length of short segments of a ladder, as it were telescopically, up against the perpendicular face of the shaft, and climbing simultaneously upon the lengthening out ladder. A number of ladders of 15′ length are in the first instance prepared, which are identical with each other in detail and form, and which are so fashioned that the bottom of any one ladder can be dropped into sockets provided at the top of any of the rest. The sides of each segment are pivots at the bottom and sockets at the top. There are also standards or pegs about 8″ long projecting out from one face of each segment, which serve the purpose of keeping it off the brickwork when it is fixed and by this means providing a secure foothold and handhold.

The first step in the erection of the ladder consists in placing one of the sections standing perpendicularly upon the ground against the bottom of the chimney. A workman then drives an iron dog or holdfast firmly into the brickwork 1′ up from the bottom of the ladder and 1′ down from its top. These holdfasts are of a hooked form, so that they can each be made to clamp one of the rungs of the ladder when they are driven home upon it into the brickwork. The segment of the ladder is firmly attached to the shaft of the chimney when this has been accomplished.

When one section of the ladder has been attached in this way a free ladder is sloped against it and the climber then ascends upon this until he can reach a foot above the top of the fixed segment. He there drives in a holdfast and attaches to it a pulley and block, so that one end of the rope reeved into the pulley can be brought half down a second loose section of the ladder, placed perpendicularly and side by side with the first. The rope is there fastened at midway height, and by means of the block the second section of the ladder is hauled up by men standing upon the ground until it projects half-ladder height above the section No. 1. In that position it is temporarily lashed to the fixed section, rung to rung, so that the climber can mount to its top and drive a holdfast into the brickwork 1′ above its upper extremity. He then shifts the pulley and block to this upper holdfast and descends to the ground. Section 2, still attached to the rope at its middle part, is then hoisted up to its full height above section 1. The climber, following its ascent, next inserts the pivots of its sides into the sockets at the top of section No. 1, mounts upon its steps as, still held by the pulley, it leans against the chimney, drives home two hooked holdfasts, clamping its rungs to the chimney, near the bottom and near the top ; and this having been done the second section remains fixed in continuation of the first, and the ladder attached to the brickwork has thus grown from 15′ to 30′ in height. The climber is then able to mount to its top, 30′ up the chimney, and extending his arm about 1′ higher upon the brickwork, drives in there the holdfast which becomes the *point d'appui* for the hauling up a third section of the ladder, first half its length and then full height above the second segment, so that it can be in its turn pivoted into the sockets. The third section, in doing this, is handled in every essential particular like the first, pulled half-ladder high, temporarily lashed to the topmost rungs of the fixed ladder, then lifted to its full height, pivoted into the sockets of the fixed ladder there and clamped firmly to the brickwork, and the fixed ladder has grown to a length of 45′, by the junction of three segments of 15′ each. This process is afterwards repeated with other sections of the ladder again and again, half lengths at a time, until a perpendicular path has been laid from the bottom to the top of the chimney. A chimney 255′ high, it will be observed, requires seventeen sections of the ladder to reach to its top.

The essential points in this ingenious process are : (1) The temporary lashing of each section of the ladder when it is half way up, so that the climber can get safely to the top, as it is held still attached to the pulley, and fix a fresh block above its upper extremity for the accomplishment of the second half of the hoist; (2) the joining of the sections by appropriate sockets as each one is placed in position upon the one beneath; and (3) the fixing of each section, when it is once lifted into its place, by the holdfasts driven into the brickwork of the chimney. The ladder virtually creeps up to the top of the chimney, joint above joint. The process is so easily performed by practised hands that the highest chimneys are scaled in brief intervals of time.

The chimney at the Abbey Mills Pumping Station, near Stratford, 230' high, was laddered its entire height in three hours and a half by this method.

VENTILATING SHAFTS.

VENTILATION OF SEWERS BY MEANS OF TALL CHIMNEY SHAFTS.

This means of ventilation has been used, where permission could be obtained from the owners of shafts, at

CARLISLE.	BOLTON.	SUNDERLAND.
COVENTRY.	BURSLEM.	TYNEMOUTH.
BURTON-ON-TRENT.	HALIFAX.	WHITEHAVEN.
BIRKENHEAD.	HYDE.	WIDNESS.
BLACKBURN.	NEWCASTLE.	WOLVERHAMPTON.
BOLTON-CUM-LINACRE.	ROTHERHAM.	YORK.

Carlisle.—The Carlisle sewers, since their construction in 1855, have been ventilated by tall factory chimneys. This city was one of the first to take advantage of this help to sewer ventilation, and there are about thirty tall shafts connected with the sewers.

Messrs. P. Dixon & Sons, of Shaddongate and West Tower Streets, were the first to allow the experiment to be made, on the understanding that if it was found to be injurious to the works the Carlisle authorities would cut off the connection; this, however, was not required to be done. The sewers in the neighbourhood of their tall chimneys are well ventilated, the current of air passing through one of the ventilators connected to the Shaddongate shaft, 300' high, having a velocity of 50 miles per hour, the pressure of air at the base of the chimney being equal to a column of water $1\frac{3}{10}$ of an inch in height. From experiments made by Mr. H. U. McKie, City Surveyor, Carlisle, it was found the sewers were perceptibly ventilated for a radius of 400 yards, equal to an area of 502,656 square yards, or over 103 acres, and if the system of sewers and house drains had been laid out and executed with a view of being ventilated by this shaft the surveyor had no doubt the radius could have been considerably extended.

Leicester.—No. 25 chimney shafts have been connected to the sewers of this town, and the Corporation are obtaining permission from manufacturers whenever they can to extend the system.

Sunderland.—No. 9 shafts are connected to the sewers here, and the surveyor says they are *not* a success.

Great Yarmouth.—No. 5 shafts, 50' high, have been specially built in connection with the main sewers to act as ventilators.

Coventry.—No. 15 shafts are here connected to the sewers of the town.

York.—No. 3 shafts are here utilized as ventilators.

Hereford.—No. 1 shaft only connected to sewers and the effect is quite local, the few ventilators adjoining invariably act as *down* cast shafts and the chimney as an *up* cast.

Blackburn.—In one case only is a chimney connected to aid the sewer ventilation.

Bolton.—A limited number of shafts have been utilized as sewer ventilators in this town, and with good results.

Mr. E. Buckham, Borough Surveyor, Ipswich, does not share in the fear that damage is likely to arise from explosions caused by gas leaking into the sewers, and thence travelling to the chimneys; he has not heard of such an accident and thinks the possibility of it occurring most remote. The fact that sewers are only affected by these shafts to a limited extent is, in his opinion, rather in favour of their use than otherwise, because where the exhaust is too powerful there is a probability of the traps of the house drains becoming unsealed.

GENERAL.

AN AWKWARD DILEMMA.

In September, 1872, a chimney was being erected at Messrs. Smith's Works, Alyth, Scotland, and had reached a height of 100'. One evening, when the builder was about to descend, he discovered that the rope by which he was to reach the ground had fallen from its fastening. After various plans had been suggested and partially tried to rescue the isolated man, such as throwing up a stone with a cord attached and building a temporary wooden stair inside, the happy thought occurred to the builder of taking off one of his stockings and running down the yarn so as to reach the ground. The scheme succeeded. On the end of the woollen yarn reaching the base a small cord was attached. After this had been hauled to the top an ordinary rope was secured to the end of the cord and the rope drawn to the top, by which means the builder descended in safety to the great satisfaction of the hundreds of spectators who had assembled at the foot of the shaft.

A NOVEL DINNER PARTY.

Nottingham.—At the Stanton Iron Works Company, near Nottingham, a chimney 190' high, 24' across the cap and 13' 9" across outlet, was erected in 1874. When the shaft was near its completion forty-seven of the workmen were entertained to a hot dinner at the top. Three young ladies also ascended to lay the cloth and wait upon the guests during their aerial banquet. Considerable interest was occasioned by so many people dining at such an elevation. The cap was cast at the Company's works and weighed 15 tons. The shaft contained 420,000 bricks.

Farncombe & Co., Printers, Lewes and Eastbourne.

Plate 1

TALL CHIMNEY CONSTRUCTION (R M & F J Bancroft)

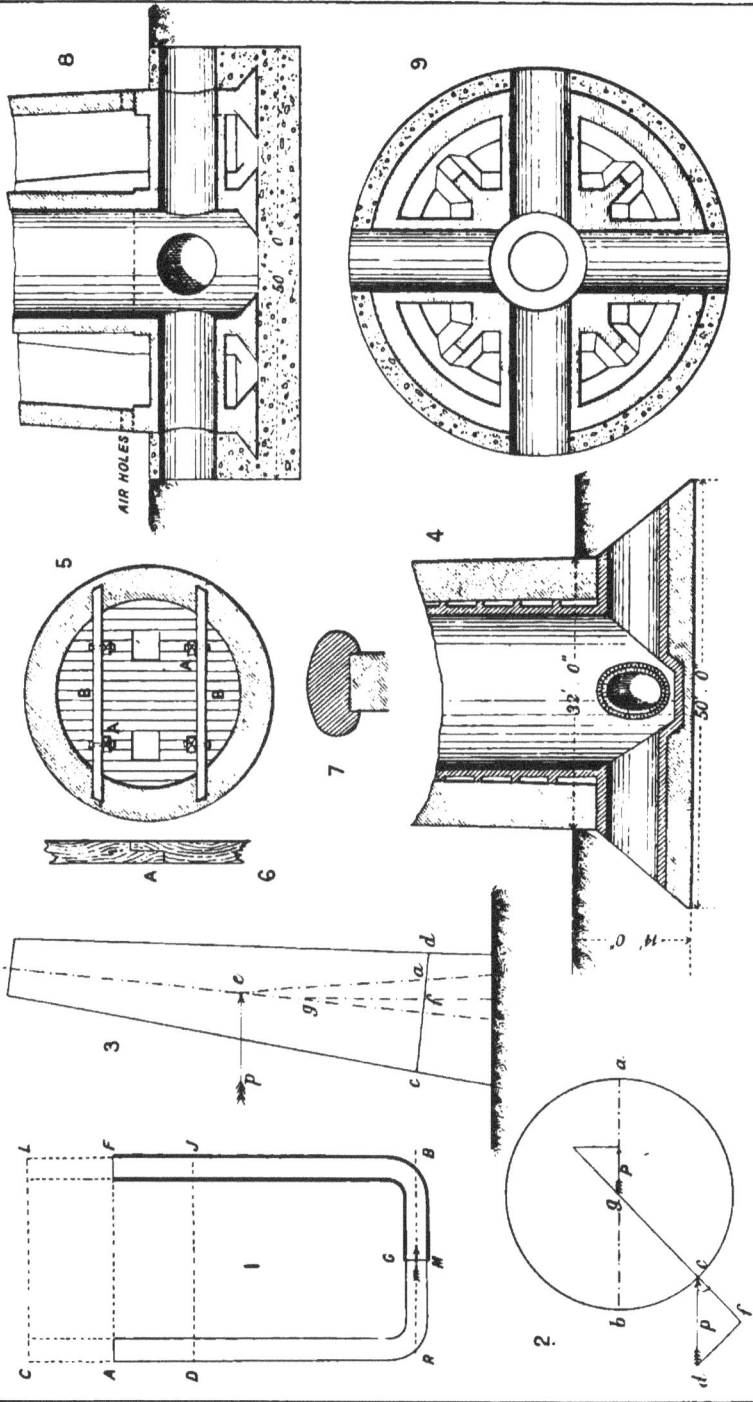

AIR HOLES

Farncombe & Co., Lith, Lewes & Eastbourne

Plate 2

TALL CHIMNEY CONSTRUCTION (R M & F J Bancroft)

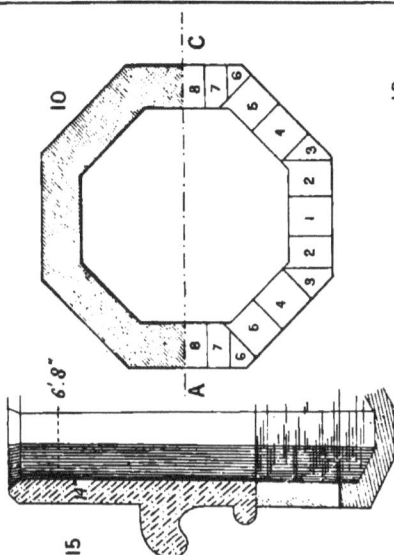

10

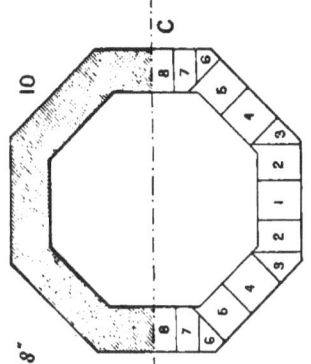

6·8″

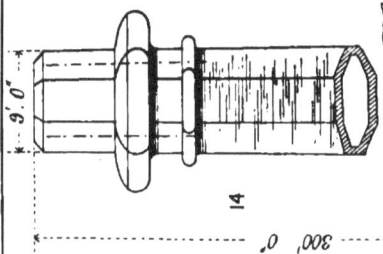

15

9′. 0″

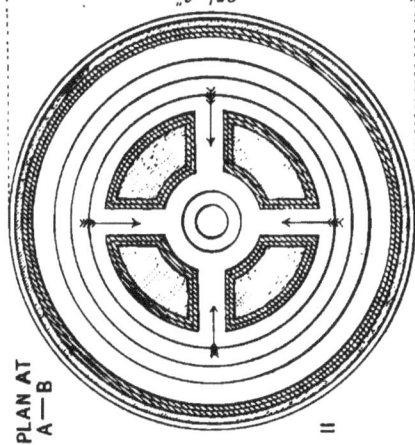

PLAN AT
A—B

11

35′. 6″

300′. 0″

14

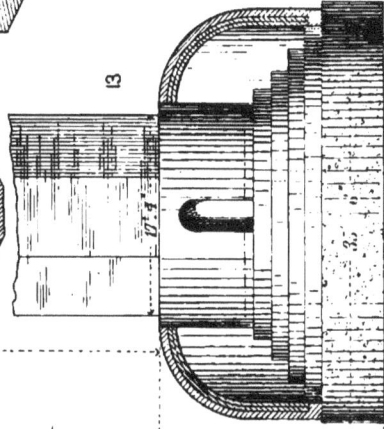

13

6′. 10″

3′. 0″

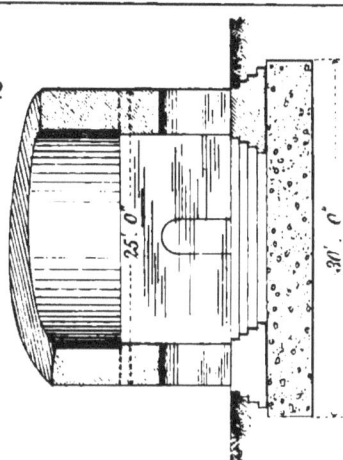

16

25′. 0″

30′. 0″

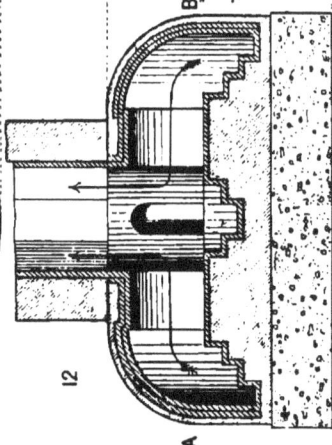

12

20′. 0″

A

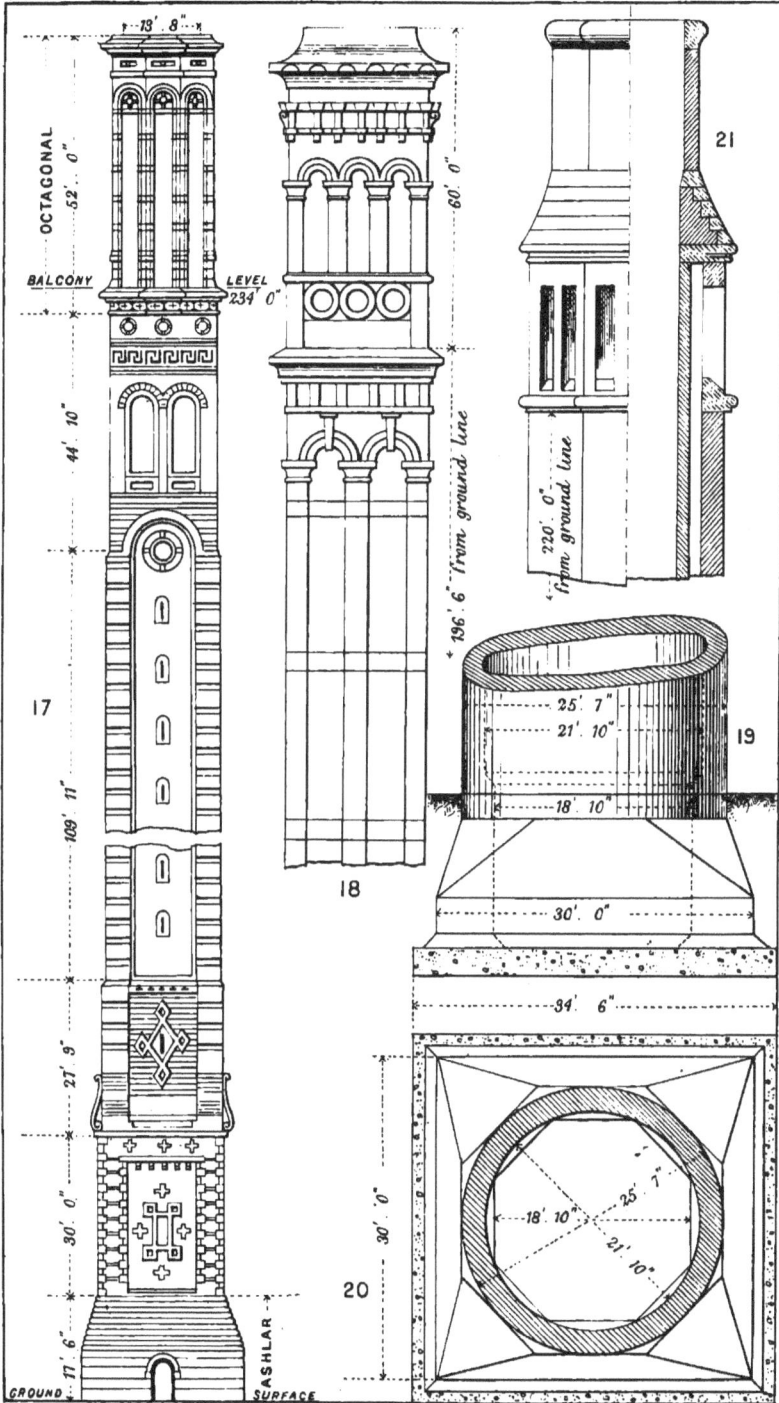

OCTAGONAL

13' 8"

52' 0"

BALCONY

BALCONY LEVEL
234' 0"

44' 10"

17

109' 11"

27' 9"

30' 0"

11' 6"

GROUND

CASHLAR
SURFACE

60' 0"

18

196' 6" from ground line

220' 0' from ground line

21

19

25' 7"
21' 10"
18' 10"

30' 0"

34' 6"

30' 0"

20

18' 10"
25' 7"
21' 10"

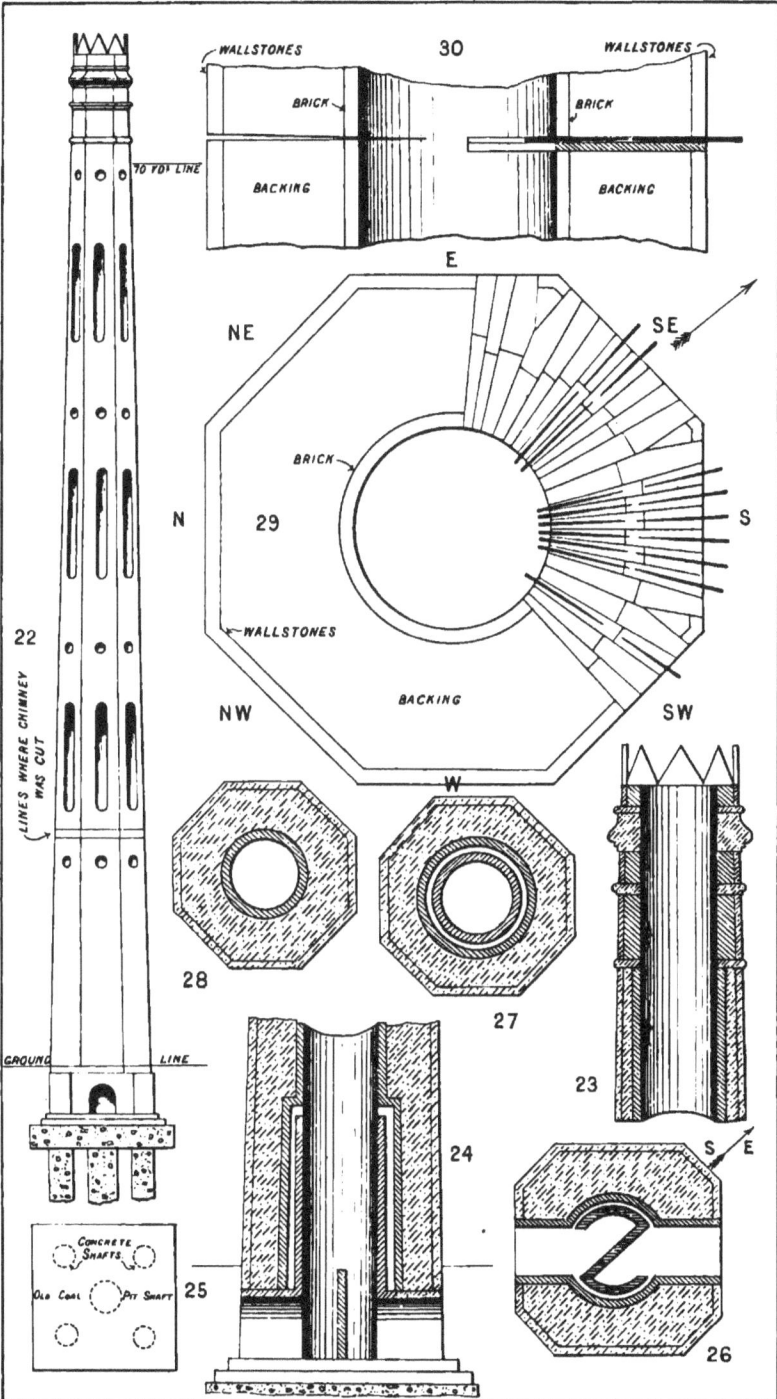

.

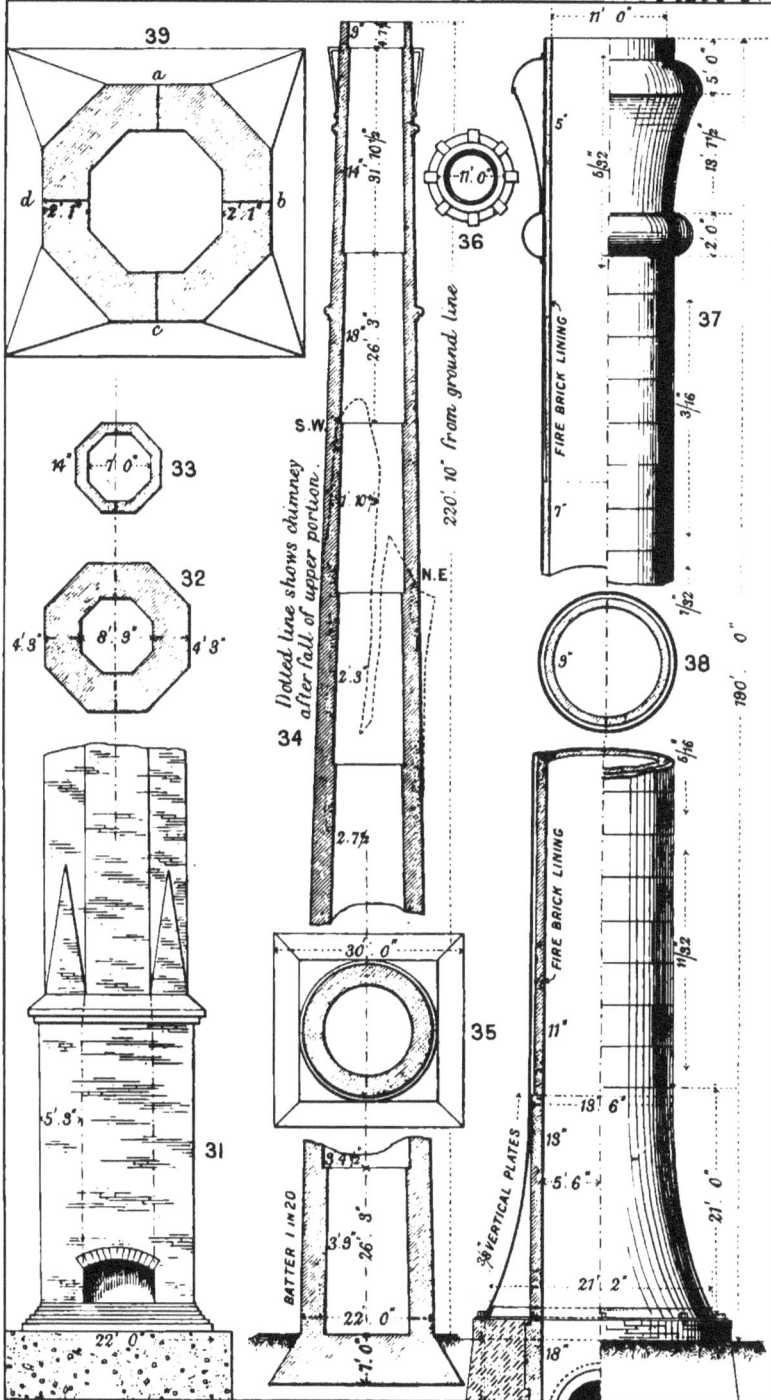

39

a

d   2'.1"   2'.1"   b

c

33   4"   1' 0"

32   4' 3"   8" 9"   4' 3"

31   5' 3"   22' 0"

Dotted line shows chimney after fall of upper portion.

220' 10" from ground line

9"

91' 10½"

14"

19' 3"

26' 3"

S.W.

1' 0½"

N.E.

2' 3"

2' 7½"

34

30' 0"

35

BATTER 1 IN 20

3' 4½"

26' 3"

3' 9"

22' 0"

1' 0"

36   11' 0"

37   11' 0"

15' 0"

13' 7½"

5"

6/32"

2' 0"

3/16"

FIRE BRICK LINING

7"

7/32"

0' 0"

38   9"

180' 0"

FIRE BRICK LINING

5/16"

11/32"

11"

13' 6"

13"

5' 6"

VERTICAL PLATES

21' 0"

27' 2"

18"

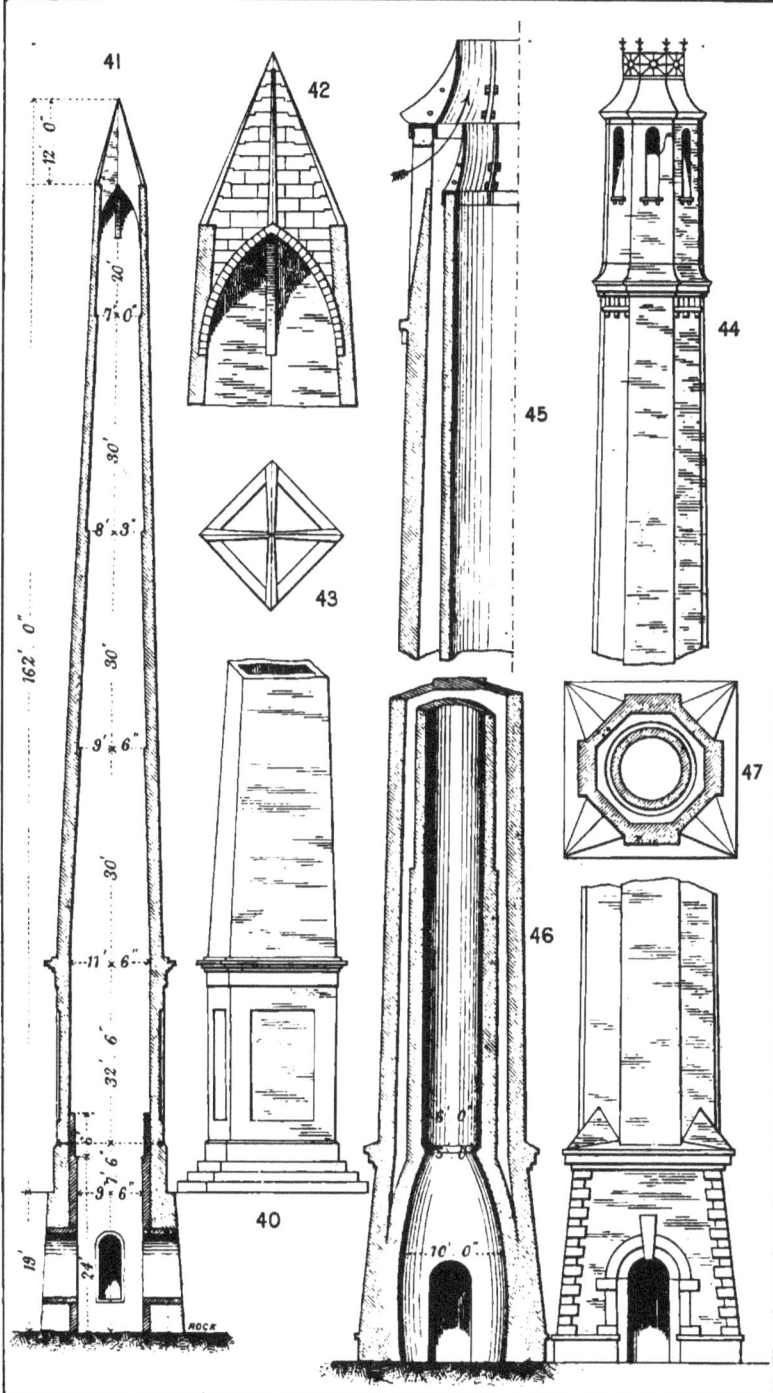

41

12' 0"

20'
7' 0"

30'
8' 3"

162' 0"

30'
9' 6"

30'
11' 6"

32' 6"
4' 6"
9' 6"

19'
24'

ROCK

42

43

40

45

44

47

46

5' 0"

10' 0"

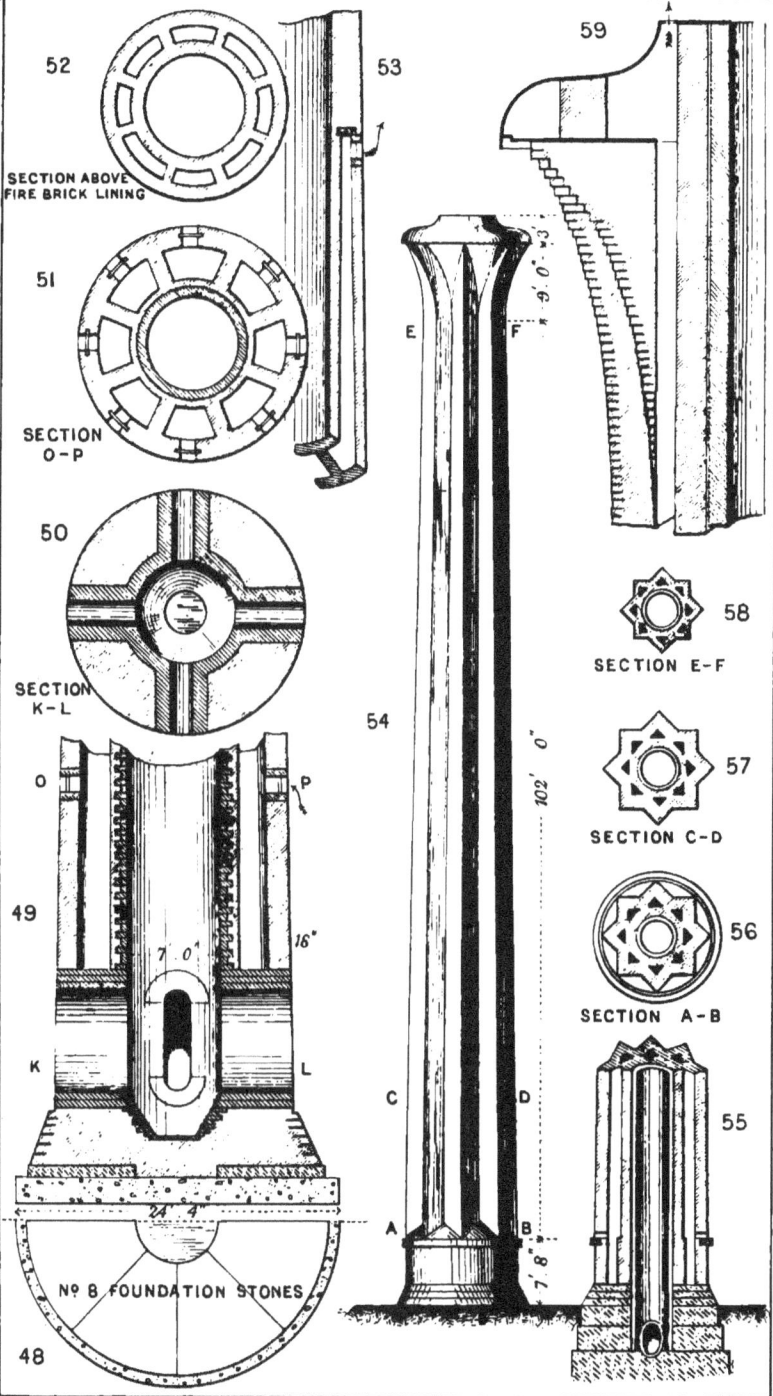

52

53

59

SECTION ABOVE
FIRE BRICK LINING

51

E   F

SECTION
O-P

58

SECTION E-F

50

54

57

SECTION C-D

SECTION
K-L

56

SECTION A-B

O   P

49

16"

7' 0"

K   L

55

C   D

24' 4"

A   B

Nº 8 FOUNDATION STONES

48

102' 0"

9' 6" 3"

7' 8"

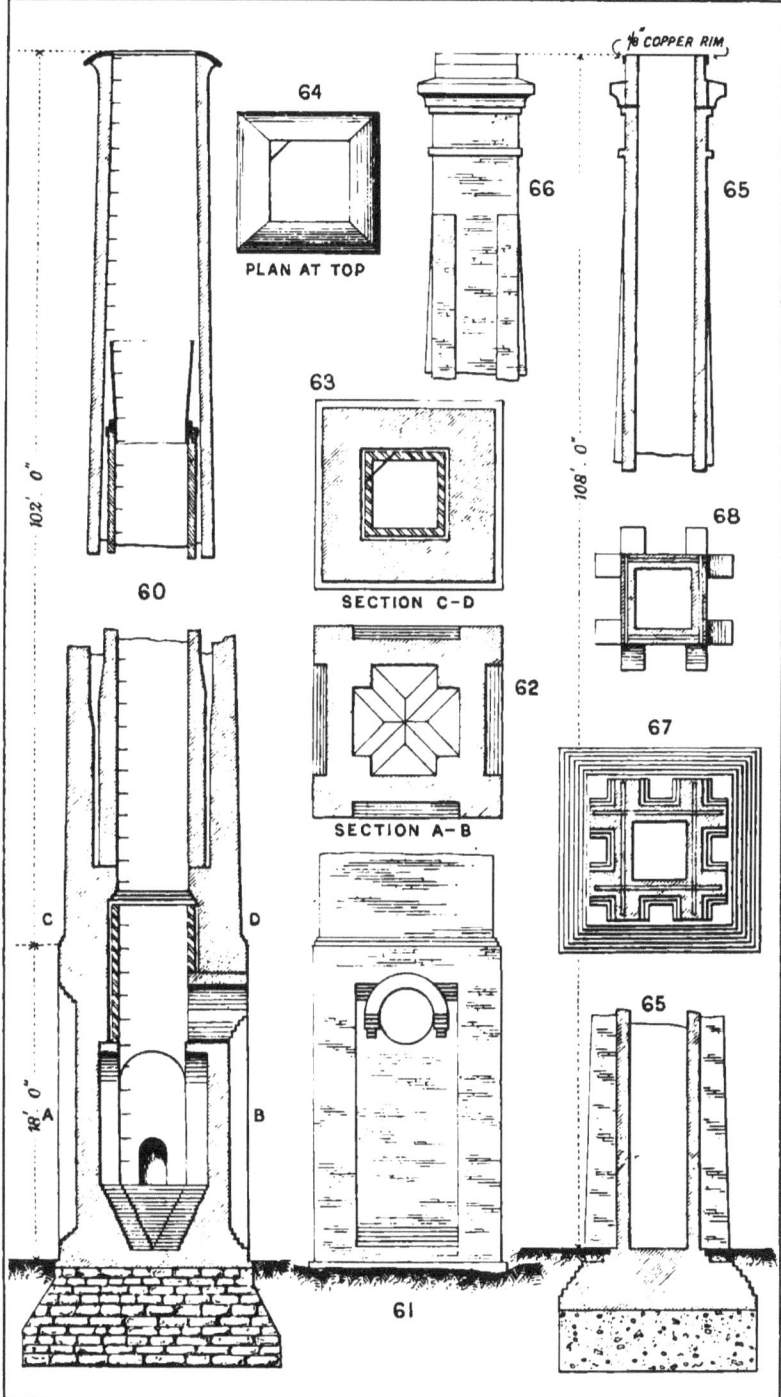

64

PLAN AT TOP

66

65

⅜" COPPER RIM

102' 0"

60

63

SECTION C-D

108' 0"

68

62

67

SECTION A-B

C   D

A   B

18' 0"

65

61

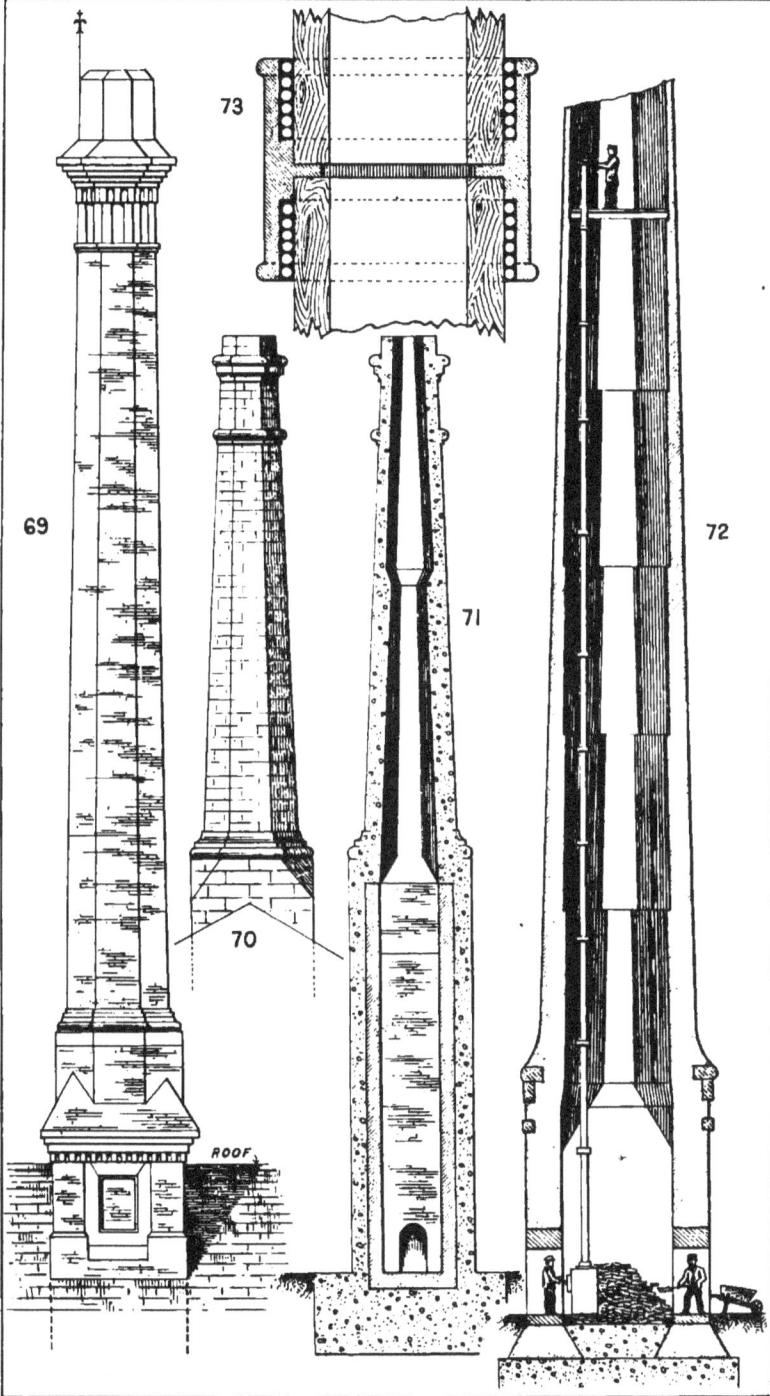

69

73

70

71

72

ROOF

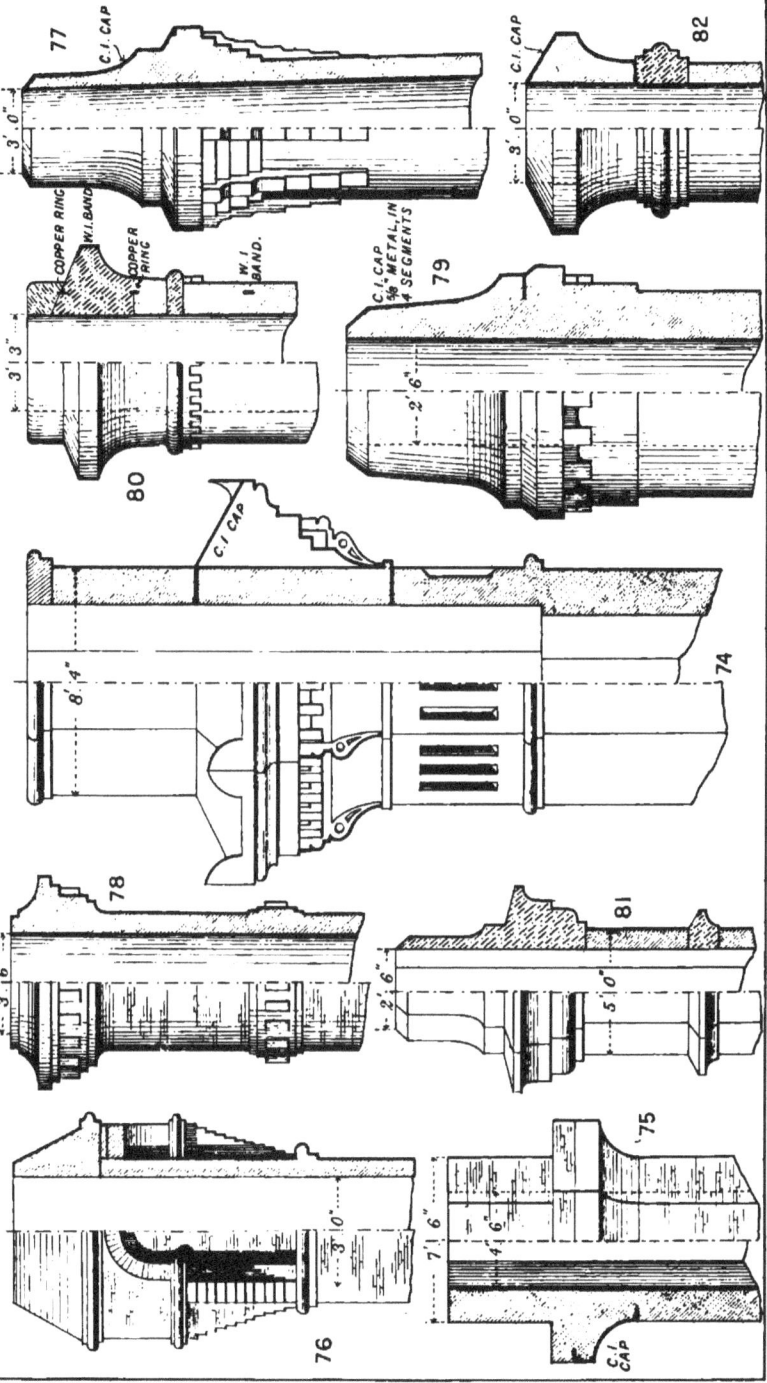

Plate 10.

TALL CHIMNEY CONSTRUCTION. (R. M. & F. J. Bancroft.

www.ingramcontent.com/pod-product-compliance
Lightning Source LLC
Chambersburg PA
CBHW021811190326
41518CB00007B/548